Making   the   Corn   Belt

MIDWESTERN HISTORY AND CULTURE

General Editors

James H. Madison and Thomas J. Schlereth

MAKING THE CORN BELT

A Geographical History of Middle-Western Agriculture

John C. Hudson

Indiana University Press

Bloomington & Indianapolis

Library of Congress Cataloging-in-Publication Data

Hudson, John C.
Making the corn belt : a geographical history of
middle-western agriculture / John C. Hudson.
p.   cm.—(Midwestern history and culture)
Includes bibliographical references and index.
ISBN 0-253-32832-2 (cloth)
1. Middle West—Geography.   2. Corn—Middle West.
I. Title.   II. Series.
F351.H858   1994
917.8—dc20
93-35723

1 2 3 4 5 00 99 98 97 96 95 94

Contents

Preface

THIS BOOK HAS been a long time in the making. More than thirty years ago it was my good fortune to encounter some exceptional teachers at the University of Wisconsin who encouraged an interested undergraduate to enroll in their advanced courses. I learned agricultural history from Morton Rothstein, economic and agricultural geography from John Alexander and Clarence Olmstead, and rural demography from Glen Fuguitt. Two extremely valuable summers spent at a desk next to Calvin Beale's in the Farm Population Branch, U.S. Department of Agriculture, permitted me to absorb as much as I could from the best tutor on rural America anyone could have had.

Graduate work at the University of Iowa in the mid-1960s brought me in contact with the reigning expert on Corn Belt geography, the late Harold H. McCarty; with a physical geographer having intimate knowledge of the middle-western landscape, Neil Salisbury; and with Lawrence Brown, who encouraged my interests in, of all things, a mathematical theory of rural settlement location. It was in Iowa City that I first learned about the "hawg," distinctly different from our "hahg" in southern Wisconsin.

The heady days of logical positivism having passed, and lacking the philosophical bent of my contemporaries, I chose instead to return to an earlier fork in the road, to pursue, in Gunnar Olsson's apt phrase, the study of "pigs and railroads." A visiting appointment at Berkeley put me in touch with two natural-born geographers of the old school, Jay Vance and Jim Parsons, and permitted me, finally, to meet the greatest American geographer, Carl Sauer. I knew I was on the right track when they proved friendly. When I returned to the Middle West, a couple of mavericks, Fraser Hart and Cotton Mather, both astute observers of the land, became my close friends.

The better part of a decade spent reading and editing manuscripts for the *Annals* made me further appreciate that, within the flurry of new directions in geography, there was a core, a comprehensive approach to understanding the world we live in, which bears the old-fashioned, textbookish label "regional geography," but which nonetheless justifies setting this whole branch of learning apart as its own discipline. Although my subject is one on which probably four-fifths of the nontechnical, scholarly literature has been contributed by historians, those who know geographers will find many of our names in the endnotes at the back of this book.

More recently, *Making the Corn Belt* originated in efforts to better understand the role of cultural factors in the emergence of the Corn Belt as a major American region. My initial approach was to map the origins of middle-western frontier populations back to the formative "hearth" areas of the eastern seaboard that have long been recognized as having exerted disparate influences on settlement west of the Appalachians. The mapping exercise provided useful background, although it soon became clear that the emergence of the Corn Belt involved far more than the transfer of agricultural practices from one region to another.

Once settlement chronology was unraveled into the various streams of westward migration, it became necessary to explain the new agricultural strategies these migrants evolved after they settled in the Ohio Valley and then to account for the evolution of Corn Belt agriculture as it spread still farther to the west in subsequent generations. Histories of crops and livestock were assumed to be of importance, although, especially in the case of corn, I did not anticipate that plant genetics and human migration would interact in a manner that favored regional agricultural specialization. The research thus led quickly from geography and history to archaeology and the natural sciences.

A Guggenheim year in 1988 took me first to the Miami and Scioto valleys of Ohio, where I had time to absorb the similarities and differences between these two important early centers of the Corn Belt, then to the High Plains, where unbelievable corn yields were being obtained under irrigation sprinklers. Subsequent field work across the Pennyroyal and the Bluegrass, Little Dixie and the Grand Prairie, the Des Moines lobe and the Platte Valley, helped bring the study into focus. Delving further into the past, I received the generous and kind assistance of my Northwestern colleague, archaeologist Jim Brown, whose pre-European version of the Middle West makes so many important contributions to what follows.

Of the many others who have helped me directly or indirectly in this study I would especially like to thank Ron, Vicki, and Marc Blakeman, John Borchert, Chuck Bussing, Michael Conzen and Kathleen Neils Conzen, Don Dahmann, Arlin Fentem, Russell Graham, Fraser Hart, Lowell Hill, John Jakle, Terry Jordan, Doug and Brenda Magnus, John Miller, John Rehder, Gil Stein, Mike Turner, Frank Yoder, and Wilbur Zelinsky. Jim Madison and Tom Schlereth, editors of the Midwest History and Culture Series, provided friendly support throughout the project. The financial support of the William and Marion Haas Fund, Northwestern University, is gratefully acknowledged.

I would like to thank the University of Michigan, Museum of Anthropology, for permission to reproduce figure 10, which appeared in Walton C. Galinat, "Domestication and Diffusion of Maize," in *Prehistoric Food Production in North America*, edited by Richard I. Ford, Anthropological Papers, 75 (Ann Arbor: University of Michigan Museum of Anthropology, 1985), 245–

78; and Michigan State University Press for permission to reproduce figures 11 and 12, which appeared in Henry A. Wallace and William L. Brown, *Corn and Its Early Fathers* (Chicago: Lakeside Press and Michigan State University Press, 1956).

The maps and photographs appearing in this book are the work of the author. Maps that include county boundaries for various census years use the base maps found in Thomas D. Rabenhorst and Carville V. Earle, *Historical U.S. County Outline Map Collection, 1840–1980* (Baltimore: Department of Geography, University of Maryland-Baltimore County, 1984). In figures 2 and 13 geological divisions are taken from United States Geological Survey, *Geologic Map of the United States* (Washington, D.C., 1974); vegetation boundaries in figures 2 and 7 are generalized from A. W. Kuchler, *Potential Natural Vegetation of the United States* (New York: American Geographical Society, 1964). Relief features shown in figures 2 and 13 are based on Edwin H. Hammond, *Classes of Land-Surface Form in the Forty-Eight States, U.S.A.* (Washington, D.C.: Association of American Geographers, 1964).

Librarians at the Steenbock Library of Agriculture and Life Sciences and the Wisconsin State Historical Society Library at the University of Wisconsin, Madison; the Seeley G. Mudd Science and Engineering Library and the Transportation Center Library at Northwestern University; the Kansas State Historical Society, Topeka; the Iowa State Historical Society, Des Moines; the Ross County Historical Society, Chillicothe; and the No Man's Land Historical Museum, Goodwell, Oklahoma, provided invaluable assistance. To Anne and Jane, and to Debby, go my love and affection and my sincere thanks for their support.

Making   the   Corn   Belt

1

Corn Belt Geography

POPULAR PERCEPTION THAT a "Corn Belt" stretches across the midsection of the United States is more than a century old. The label gained acceptance even before the present conception of "Middle West" emerged. According to William Warntz, references to this Corn Belt began to appear in popular literature in the early 1880s. James R. Shortridge's study of regional name usage shows that, until the early twentieth century, Middle West had several conflicting definitions. Its current application to states of the upper Mississippi Valley did not appear until 1912.[1] The lag is understandable, given the constant emergence of new "wests" as the frontier expanded.

Although corn/livestock agriculture moved west with the frontier, the inner core of today's Corn Belt was established by 1850. It has remained a fairly stable region, even as its margins have fluctuated with the weather, technological changes in farming, and shifts in government agricultural policy that have favored expansion into, or retreat from, a large territory surrounding the established core. The Corn Belt's significance has been doubted less often than it has been embellished. Writing in the mid-1950s, Haystead and Fite called it "The Corn-Soy Belt: Feedbag of Democracy."[2]

Oliver E. Baker was the first to define the Corn Belt quantitatively, in one of a series of articles on the agricultural regions of North America that he wrote in the 1920s. Baker's regions were based both on agricultural production statistics and on variables of the physical environment. He bounded the Corn Belt on the west by the 8″ summer rainfall isohyet, on the north by the 66° summer isotherm, and on the east and south primarily by the amount of corn produced relative to other crops.[3] Closely correlated with corn crops were the numbers of hogs and cattle, the markets for both, and the production of oats. In another article, "The Middle Country Where South and North Meet," he defined the zone of mixed corn and winter-wheat production south of the Corn Belt in terms of land resources. Baker's regions were modified slightly and issued by the USDA in 1950 as *Generalized Types of Farming in the United States*.[4] This remains the standard regionalization used to this day, in spite of the many changes that have taken place since. The "official" Corn Belt of 1950 can be compared with statistics of corn production at that time and with Baker's earlier regional limits (fig. 1).

Perhaps because Baker worked for the United States Department of Agriculture, his Corn Belt was also drawn somewhat optimistically, suggesting

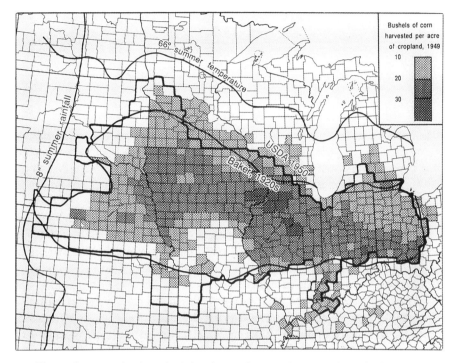

Fig. 1. Corn production (bushels of corn harvested per total cropland acres, 1949) compared with O. E. Baker's earlier delineation of the Corn Belt and the U.S. Department of Agriculture's official designation of the region published in 1950. Baker suggested that the 60° summer temperature and the 8″ summer rainfall lines would form physical limits on the Corn Belt. Corn production was shifting northward at mid-century.

a belief that the physical environment would be the limiting factor that shaped the region.[5] An optimal setting for the production of feed grains and livestock was interpreted as the foundation for farm prosperity in the Corn Belt. Baker noted the correspondence between this region and the concentration of rural telephones and automobiles on farms and with high values of farm property. High rates of farm tenancy also were correlated with the Corn Belt but were left unexplained. O. E. Baker had enormous insights into the geography of American agriculture, yet, in the fashion of that time, his regions were synchronic; he allowed little role for human cultural practices, operating over time, in creating them.

In an interesting but now largely forgotten paper published in 1963, Joseph E. Spencer and Ronald Horvath posed the question, How does an agricultural region originate? Beyond obvious factors of the physical environment, they found that human decisions were of primary importance. "Taken from the point of view of psychological aspects, social process, and cultural traditions of the settlers," they wrote, "the Corn Belt can be regarded as the landscape expression of a farming 'mentality.' " They concluded that "the working of a cultural process, in an evolutionary way" had produced the region. It was not "a gift from a pantheon of gods [nor] . . . a widespread and purposeful copying of an efficient model known to all settlers."[6] Spencer, like other cultural geographers, regarded history as process and the landscape as its product.

John Fraser Hart's *The Land That Feeds Us* treats these landscape expressions of farming mentality on a region-by-region basis.[7] Unlike the mono-cropping practices of the traditional cotton and wheat belts or the specialized routines of dairying, the middle-western Corn Belt developed as a mixed system of crop (feed grains) and livestock production. Hart regards south-eastern Pennsylvania as the "first seedbed" of this system, which, in turn, was strongly influenced by European crop rotation patterns and strategies for livestock production. He identifies the Miami Valley of Ohio as the second seedbed because its agriculture was influenced by that of southeastern Pennsylvania. Diffusion of these ideas across the Middle West produced the belt of corn/livestock farming that has endured for generations.

Baker's statistical regions, Spencer and Horvath's notions about how such regions originate, and Hart's invocation of farming habits to explain the contemporary rural landscape all suggest the authority of tradition in shaping an agricultural region. Even so it is difficult to find a Corn "Belt" in the United States by 1840, the first census in which statistics of agricultural production were reported. In 1840 there was no wide swath of counties stretching across several states where corn/livestock production dominated other forms of agriculture and where a clear regional pattern of surplus production involving numerous contiguous counties could be identified.

There were, rather, five islands of good agricultural land west of the Appalachians that supported the beginnings of intensive farming: the Scioto and Miami valleys of Ohio, the Bluegrass of Kentucky, the Nashville Basin of Tennessee, and the Pennyroyal Plateau along and north of the Kentucky-Tennessee border (fig. 2). These were the long-used and much improved lands of the native inhabitants of North America, lately occupied or used as hunting grounds by the Shawnee and other tribes, where the first white settlers penetrating beyond the Appalachians saw great stretches of land that looked ready for the plow immediately.

Settlers came from all the eastern seaboard states, but there was a distinct geographical bias in migrant selection. Trans-Appalachian immigrants were derived largely from the Piedmont and Great Valley, ranging from Pennsylvania through Virginia to the Carolinas, the eastern portion of the cultural region known as the Upland South.[8] Like all settlers heading west Upland

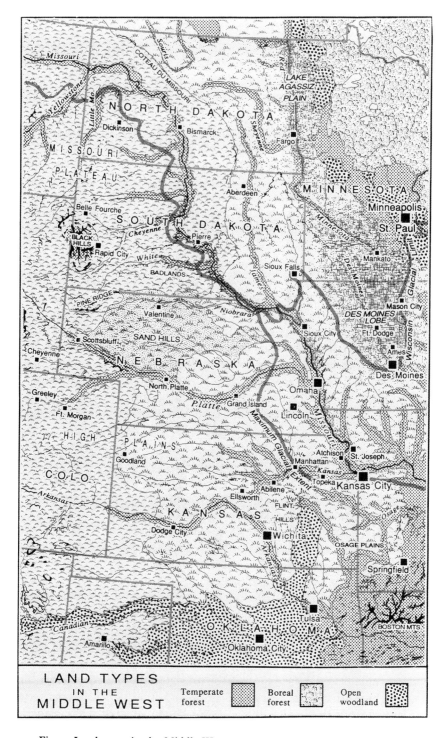

Fig. 2. Land types in the Middle West.

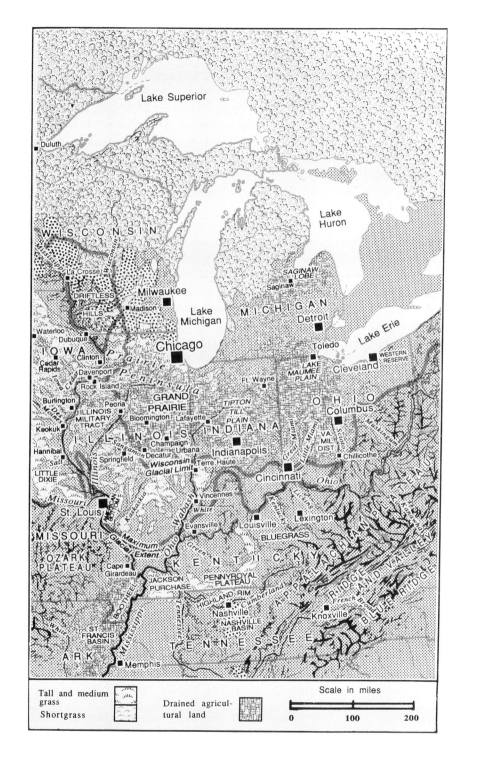

Tall and medium grass
Shortgrass
Drained agricultural land

Scale in miles

0   100   200

Southerners brought with them a store of agricultural knowledge and as many head of stock and sacks of grain as their circumstances and resources allowed. Although most of these immigrants were of English, Scotch-Irish, or German background, they carried with them, as part of what is sometimes called "cultural baggage," an array of habits derived from their immediate pasts that reflected a far wider variety of ethnic influences than their own baggage would be expected to contain.

They built cabins, barns, corn cribs, smokehouses, and a variety of other small buildings in the fashion of log pens, following the practices brought to North America by the Savo-Karelian Finns who came to New Sweden on the Delaware River in the 1640s. Their cattle and swine included breeds that had come to the New World with the Spanish years before English colonies were planted. The long-snouted razorback hog that de Soto's expedition brought from Extremadura had subsequently adopted every human group with which it came into contact; rare was the pioneer crossing the Appalachians who was unaccustomed to dining on its flesh. The seeds of European grains and pasture grasses were carried in. And, most distinctively, this group of migrants threading the gaps leading west from the Great Valley carried sacks of Dent corn, the high-yielding Mexican race that was probably introduced to the southeastern United States by the sixteenth-century Spanish.[9]

In these respects the settlers who established farms in the five islands were no different from their counterparts from the same origins who took lands of lesser quality in Tennessee, Kentucky, and southern Ohio. Good land was put into production more rapidly, however, produced larger crops of corn, and thus attracted further investment. Razorbacks were eliminated by crosses made on a variety of English swine breeds imported to the Miami Valley. Blooded cattle were imported from England to the Bluegrass and the Scioto Valley by men who had organized an efficient system of cattle feeding in (West) Virginia and then moved west to expand it.[10] Large corn crops stimulated interest in breeding animals that would consume more of the crop, just as the growing market for beef, pork, and lard encouraged greater corn production. These various attempts to intensify agriculture, to increase both the quantity and the quality of production, were what set the five islands apart from the rest of the Ohio Valley, and thus it is here that the nascent region can first be identified.

The five islands of increasing agricultural intensity of the mid-nineteenth century shared little in common when it came to their respective modes of constructing a community, however. The Miami Valley was the stronghold of small farmers and democratic equality; productivity per acre was high, and tenancy was the exception. The Scioto Valley was initially dominated by cattle feeders from the South Branch of the Potomac River, while land in the Virginia Military District was held in large tracts and often rented to tenants;

the region's roots were Southern even though its location was Northern. The Kentucky Bluegrass embraced a scaled-down version of Virginia's slave economy; its agriculture included labor-intensive crops such as tobacco and hemp, yet its corn/livestock production rivaled the Miami and Scioto valleys' until the Civil War. The Pennyroyal Plateau had a mixture of crops similar to the Bluegrass, although with a lesser emphasis on plantation-style agriculture. More than thirty-five thousand slaves were held in the Nashville Basin in 1839, where cotton was a cash crop at the same time that corn and swine production were increasing. The Nashville Basin produced the most corn of any of the five islands in 1839, and its hogs were largely responsible for Tennessee's top rank in swine production at that time.

Clearly there can be no apposition of a particular social system or a single community mentality with corn/livestock agriculture as it developed west of the Appalachians in the first half of the nineteenth century. Yet despite lack of a shared philosophy in other matters, those who moved west from the five islands did share farming practices in common. Kentuckians seeking to avoid slavery moved to the Sangamon country and the Wabash Valley; their Bluegrass neighbors who wished to extend the territorial scope of slavery went to Little Dixie. From these extensions the Corn Belt fanned out to the west, continuing to grow until it reached the edge of the Great Plains in the 1880s. Good land that produced large corn crops, more than anything else, would map the westward spread of the Corn Belt.

The essence of Corn Belt agriculture is the practice of fattening hogs and beef cattle on corn.[11] Not until cash-corn production began to emerge in eastern Illinois around 1860 did any region specialize in producing corn for a purpose other than feeding the farmer's own stock. Corn was, of course, ground for meal intended for direct human consumption, and the surplus production of some counties in the Ohio Valley was destined for whiskey distilleries, but neither of these uses can account for more than a fraction of the geographical expansion of production. Despite the common habit of planting corn and the use of the crop to fatten animals, however, there is nothing in the term "Corn Belt" that would suggest a definition of the region sufficiently precise to demarcate it on a map. There must be some definition or test that can be made to determine whether or not an area (a county, using census data) is to be included or excluded. A value of 7.5 bushels of corn per acre of improved farmland offers a reasonable threshold of corn production, one that varied only slightly over the last half of the nineteenth and early twentieth centuries, to use as the limit of a recognizable belt of corn farming.[12]

A second criterion involves the production of sufficient corn to feed the numbers of animals reported in the census statistics; a limit of 18.5 bushels of corn per head of livestock similarly offers a reasonable threshold value for livestock feeding rather than grazing. Thus, individual counties will be considered as part of the Corn Belt if their production statistics satisfy two test conditions: first, if their output of corn was at least 7.5 bushels per acre of improved farmland; and, second, if they produced at least 18.5 bushels of corn per head of hogs and cattle. The two tests are then applied to decennial census figures to produce maps of the Corn Belt for various years. These criteria are the basis of the Corn Belt maps through 1920, the last year in which the census reported statistics on improved farmland.

The first test identifies counties where corn was an important crop. One way to interpret the threshold of 7.5 bushels of corn per acre of improved farmland is to envision a farm that had one hundred acres of improved land. Subtracting, say, twenty acres of uncultivatable land remaining as permanent pasture, the remaining eighty would be used in a rotation of crops.[13] Rotation pastures were planted in an annual hay crop that provided fodder for cattle; corn was fed to cattle and hogs, and oats were grown to feed horses as well as other farm stock; a wheat crop, if grown, was usually sold for cash. If a four-year rotation, such as hay/corn/oats/wheat was used, then twenty acres of improved land would be devoted to corn each year. A rather large corn yield, 37.5 bushels per acre, would thus be required in order to produce 7.5 bushels of corn for each acre of improved land. If the wheat crop was omitted, as it was in many parts of the Corn Belt, then the rotation was typically hay/corn/corn/oats. In this case the corn yield would have to be no larger than 18.75 bushels to the acre to satisfy the first test.

The second test separates livestock feeding counties from dairying or livestock grazing counties. The practice of fattening hogs and cattle on corn was widespread by 1840, although the corn crop was insufficient in most counties to be able to count livestock as corn-fattened. The unimproved breeds of cattle and swine first found west of the Appalachians were not able to gain weight rapidly before being driven to market. To bring in cattle and hogs at heavier weights required both a large corn crop and livestock that could convert that corn into animal fat and protein. Counties that produced a fair amount of corn (satisfying the first test) but raised insufficient corn to fatten the number of animals reported were a vanguard of the Corn Belt that was to follow. They formed a fringe, a sort of "woods Corn Belt," where mast for hogs was abundant and where woodland openings or prairie for cattle were also found, but where agriculture had not intensified so much that livestock feeding was the common practice. These counties also were most likely to be those where improved breeds of cattle and hogs either had not been introduced or were unimportant as a fraction of the total number of animals. Farmers who marketed only razorback hogs or scrub cattle had little incentive to produce large amounts of feed grain.

The Corn Belt's early islands are just visible in 1840 (fig. 3). Large corn

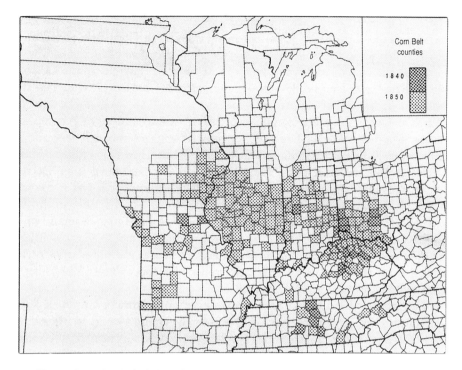

Fig. 3. Counties included in the Corn Belt, 1840 and 1850.

crops fed to swine and cattle characterized several counties in the Scioto and Miami Valleys, the Bluegrass, and the Nashville Basin. Apart from a few Indiana and Missouri counties that were growing corn crops large enough to feed the numbers of cattle and hogs reported in the 1840 census, the pattern only foreshadows what was to emerge after another decade. Thus, although the ingredients of Corn Belt agriculture—Dent corn, improved breeds of cattle and swine, and the farming skills necessary to produce fat stock—all existed by 1840 and were spreading to the west, time was needed to expand and intensify production.

By 1850 the Corn Belt had blossomed into a recognizable shape. In Indiana the Tipton Till Plain and the Wabash Valley were added; the Sangamon country, Grand Prairie, and Military Tract came into the Corn Belt in Illinois; Missouri's Little Dixie and Osage Plains also were included. The Corn Belt was a continuous region, stretching more than five hundred miles from southern Ohio to eastern Iowa. It was an expansion and, more importantly, an intensification of the 1840 pattern. The diagonal strip of counties from southwestern Ohio to eastern Iowa that formed the Corn Belt's advancing frontier was destined to remain as the heart of the region. A century later, when hybrid seed corn was introduced, these same central counties would lead the adoption process. Although many changes in Corn Belt agriculture have taken place since 1850, the core evident by that date has remained.

As a regional-cultural type, the Corn Belt is rooted mainly in the Upland South. Because these middle-western counties were settled prior to 1850 by sons and daughters of the earlier generation, who had lived in the five islands, the Corn Belt was expanding north and west as a result of transplanting an agricultural type from one region to another through migration. There were Yankee contributions, such as the Byfield hog from eastern Massachusetts and the steel plow manufactured by Vermont-born John Deere. But Yankees did not invent Corn Belt agriculture, and neither was it a common style of farming in their section of the upper Middle West until the later decades of the nineteenth century.[14] Nor even did Pennsylvanians who moved straight west into Ohio and Indiana initially establish Corn Belt agriculture in the region they dominated. Rather, it was the presence of early settlers from the five islands that determined the appearance of large corn crops and the production of fat stock.

The notion that Upland Southerners were poor farmers cannot be sustained in the face of this evidence. The reason the Grand Prairie, southeastern Iowa, the Sangamon country, and the Wabash Valley all practiced the same mode of farming was that, despite the various population sources that comprised the migration flows to those areas, all of them were settled by farmers from the five islands. The stereotypical attitude concerning Yankee/Southern differences in level of economic ambition has a secure historiographic base. According to Eric Foner, "Southern-born settlers seemed content with their status in life and appeared to lack the typical northern desire to improve their condition." In his classic regional study, *From Prairie to Corn Belt*, Allan Bogue criticized the "Yankee determinist" view of middle-western history, but the canard remains.[15] If Upland Southerners were merely "content," how could the Kentucky-born practice the same type of agriculture as the Ohio-born, and at the same level of intensity, and thereby create the Corn Belt? There is no need to invoke a later "Yankee infusion" to explain Corn Belt agriculture. The Corn Belt was established before the Yankees were, and it was the Yankees who learned, at least as much as they taught.

Corn Belt agriculture spread across central Illinois and included twenty counties of Iowa before Chicago became a factor in the region's economy. The city was peripheral, geographically as well as functionally, until a canal and several railroads connected it with outlying areas.[16] The first Corn Belt looked south, down the Mississippi River system, but that orientation disappeared once railroads entered the Mississippi Valley. The shift from river to rail transport inverted logistics and made peripheral Chicago the focal point. Before 1850 it would have been difficult to predict that a city with only a fair

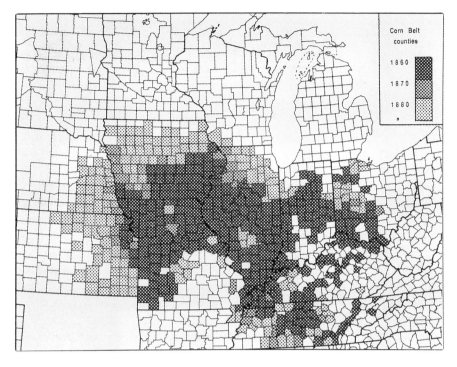

Fig. 4. Counties included in the Corn Belt, 1860, 1870, and 1880.

harbor on Lake Michigan would soon set the price of corn and butcher the livestock raised on Corn Belt farms.

Migrants from Kentucky and Tennessee settled Missouri and most of southern Illinois and Indiana. Those who took hilly, broken lands in the Ohio Valley and the Ozark fringe were unable to intensify their corn/livestock production to the levels found farther north, although their style of farming was similar. Even the glaciated counties of southern Illinois and northern Missouri took longer to develop livestock feeding economies; they were not included in the Corn Belt until 1860 (fig. 4). These counties also were among the first to diversify away from corn/livestock agriculture after 1880. Richard K. Vedder and Lowell E. Galloway described southern Illinois as "geologically disadvantaged" because it lies south of the "terminal moraine."[17] In fact, many of these lagging counties were glaciated, but in earlier ice advances. The main break in soil quality is the Wisconsinan glacial limit (fig. 2). Southern Illinois's "gray lands" were glaciated, but the soils there are poorer for a nutrient-demanding crop like corn.

There was little difference in soil fertility between glaciated and unglaciated lands of equivalent slope in the early history of the Corn Belt. John

Filson reported that some Bluegrass farmers had yields of one hundred bushels of corn to the acre in their first season, in the early 1780s.[18] Differences appeared over time because the older soils, south of successive glacial limits, were less able to hold nutrients. Unglaciated plateaus developed on calcareous limestones in Kentucky and Tennessee were high in natural fertility, but continued cultivation caused soil loss, the topsoil became thinner with use, and eventually cornfields reverted to bluegrass pastures. Corn production declined in Tennessee and Kentucky after 1880. In 1950 only five counties south of the Ohio River, surrounding the Green River in western Kentucky, were classified "Corn Belt" by the USDA.

Northeastern Kansas and the Missouri Valley of Iowa were included in the Corn Belt by 1870, and within another decade corn/livestock farming was projected west beyond the 98th meridian along the Nebraska-Kansas border. Farmers continued to test the dry limits of corn culture during the boom years of the 1880s, only to retreat in the droughts of the '90s. Baker's delimitation of the Corn Belt's western limit at the line of 8" summer rainfall may have been an attempt to anticipate the future, but dryland corn production has yet to be established that far west on a continuing basis. Irrigation has, at least temporarily, erased any western limit. Fossil groundwater pumped up from the underlying High Plains aquifer currently supports corn production in eastern Colorado, where precipitation often supplies less than half of the crop's moisture need.

Few areas suitable for grain-corn production remained untouched by commercial agriculture in 1880. The Corn Belt expanded slowly northward after that time, especially into Minnesota and Wisconsin, where established wheat and dairy farmers—mostly Yankees, Germans, and Scandinavians—began raising corn to fatten hogs. Here is a good illustration of the role of human habits in forming an agricultural region: The rapid unfolding of the Corn Belt between 1840 and 1880 was produced by the westward migration of people already accustomed to this style of agriculture. Subsequent northward movement required a change in farming practices. That, more than climatic restraints, produced the lag. Nor has the Corn Belt yet reached a northern limit. Boreal forests of the northern Middle West were not significantly invaded by cornfields until the 1970s, by which time Baker's 66° summer isotherm also had been left behind in the northward movement. Grain corn is grown successfully in southern Manitoba today, and climatic limits on corn's northward spread continue to be erased by still shorter-season, longer day-length hybrids.

Corn production in the Middle West has tripled since 1939, even after farmers had begun growing hybrid corn. Part of the increase has come through plant breeding, but bigger machinery and imported natural resources have contributed even more. Grain combines, anhydrous ammonia tanks, herbicide sprayers, center-pivot irrigation machines, gas-fired grain dryers, high-horsepower tractors, and trains of 100-ton grain hopper cars have redefined Corn Belt geography just as they have reshaped the Corn Belt farm. Heavy use of chemical fertilizers and increased reliance on technologies that burn fossil fuels have produced a Corn Belt that now extends the entire breadth of the Middle West and Great Plains: from Pennsylvania's border with Ohio to the Rocky Mountain foothills north of Denver. The Corn Belt has always fed heavily on the land supporting it, only now that land has to be replenished by fertilizers imported from outside.

It would be wrong, though, to assume that these recent changes were the first steps in removing the Corn Belt from a total reliance on nature for its productivity. The land would not have been nearly as productive in the past had it not been for two massive works of humankind, burning and drainage. Although alluvial bottomlands were the favored sites for early corn-fields, much of the Corn Belt that has developed since 1850 rests either on prairies, drained land, or both (fig. 2). Burning and drainage are destructive of existing habitats, yet neither of these simple expedients can be considered deleterious alongside complex modern-day technologies that typically generate negative side effects. Burning was the practice of aboriginal North Americans, especially Indians of the prehistoric period, whose efforts to increase bison populations caused the tallgrass prairie to expand into the eastern woodlands. Drainage came later, the work of European-Americans who became skilled in the ways of digging ditches and laying tiles to accomplish what nature, in a mere twelve thousand years since the glacial ice melted, has not yet had time to complete.

Some of the most productive land in the Corn Belt was both prairie and natural wetland. Parts of the Grand Prairie of Illinois and the Des Moines lobe of Iowa (referring to a lobe of the continental ice sheet) were wet prairies that required drainage before they could be farmed extensively (fig. 2). Their soils are still among the richest in the Corn Belt after over a century of use. The Corn Belt in central Indiana and eastern Ohio was similarly developed on deep, organic soils that had to be artificially drained before they could be reliably used for crop farming. The Corn Belt is not a single, homogeneous block of good agricultural land across the Middle West. It is a mosaic of land types, each with its own natural history and record of human use.

The role played by people in creating this land has not figured prominently in historical works.[19] Alfred Crosby has written that Europeans "always succeeded in taming whatever portion of temperate North America they wanted within a few decades and usually a good deal sooner."[20] Crosby contrasted the almost "fluid advance" in North America with the slow progression of settlement in the African *veldt*, the Siberian *taiga*, or the *sertao* of

Brazil. Crosby's conclusion arises from the examples he chose for comparison: none of his three was comparable with the quality of land in the future Corn Belt, where the advance was most fluid. But did Europeans actually tame this land? Or did their settlement advance rapidly because there was little taming that needed to be done?

The idea of "taming" nature requires assumptions about a prior wild landscape. Just as our contemporary concept of a "natural vegetation" is based largely on impressions recorded by early Euro-Americans, the notion that the pre-European landscape was wild, the product of nonhuman forces, and was in some sense waiting to be occupied, is clearly wrong. The land that was to become the Corn Belt had a long record of human modification before Anglo-Americans penetrated beyond the Appalachians. To understand the emergence of the Corn Belt requires a look at this process, one that continued nearly until the moment white settlers arrived. If these earlier human populations had made only a transitory impact on the land, then they might be treated as a well-bracketed chapter of settlement history that just happens to precede occupancy by the descendants of Europeans. It was, rather, the very presence over millennia of people who improved the landscape for a civilization that followed that accounts for the Corn Belt's rapid rise in the mid-nineteenth century. It was not a purely natural landscape that these Europeans entered, but one that had been enhanced beyond what nature alone had provided.

2

Making the Land

IT IS EASY to imagine that the Corn Belt is a gift of nature. From the Cumberland to the Kansas, from the Scioto to the James, the broad, undulating plain is blessed with rich soils, adequate moisture, and a warm growing season. The centerpiece—nearly one hundred million acres of a once-prairie grassland—is fringed by upland hardwood groves and rich bottomlands, all equally and ideally suited to the production of grain crops. One might presume that these ingredients have always been present, just waiting for the right application of human ingenuity to create a bountiful expanse of farms. If human history had begun with the arrival of white settlement around A.D. 1800, and if no modifications of the landscape had occurred for thousands of years before that, then perhaps such a view of a passive Nature, made to yield to the plow and the scythe of Man, would be credible. But facts confound belief in such an uncomplicated scenario.

For one, nature has not been a constant. The end of the Pleistocene (Ice-Age) epoch accompanied changes both swift and dramatic in their impact on the land. Only about twelve thousand years before the present (12,000 B.P.) the climate warmed, humans became established across North America, and the glaciers disappeared. The period of the last ten thousand years, known as the Holocene epoch, has seen alternating climatic shifts between cool-wet and warm-dry conditions and a nearly total change in the patterns of vegetation and soils. Thus the natural environment that now exists (in most areas, only in a theoretical state) came into its present configuration a matter of only thousands of years ago.

Other environmental changes were wrought by human beings. Two centuries of habitat alteration at the hands of a Euro-American population have been massive. But 110 centuries of earlier human habitation left an impact as well. Natural history in the Corn Belt is intimately connected with human history. Beginning with the arrival of the first big-game hunters who entered the interior of the North American continent as the Pleistocene was drawing to an end, on through the era of flourishing American Indian civilizations, down to the impoverished nature hemmed in by our dominating culture of today, there has been continuous landscape change as a result of the uses people have made of their surroundings. The distinctiveness of the Corn Belt region has both a natural and a cultural basis, for probably less than one-twentieth of the time that has passed since the glaciers disappeared has been without human witness. Nor were there many pristine environments remaining at the time of Euro-American contact that would offer a benchmark for "before" and "after" comparisons, contrasting a pure nature with its degraded, Europeanized successor. Landscape change has occurred in all periods, the result of both human activity and natural forces beyond the control of people.

The abrupt shift in the environmental record that took place when the glaciers disappeared twelve thousand years ago is the most significant time divide in the Corn Belt's prehistory. Reaching back even another two thousand years one finds the glacier still advancing sporadically, with new morainic ridges at the edges of various lobes of the Laurentide ice sheet in Ohio and Iowa.[1] The glacier continued to retreat generally northward, interrupted by occasional advances, until its great mass was reduced to nothing over mid-America. Curving moraines of glacial debris left by the wasting ice sheet punctuated rolling expanses of smooth plains (fig. 5). Huge lakes formed at the glacial margin eventually drained and evaporated to leave almost totally flat surfaces, as in the Lake Maumee Plain of Ohio or the Lake Agassiz Plain of Minnesota and North Dakota. Rivers carrying glacial meltwaters rapidly increased in size, then shrank again after the ice had melted, leaving their broad floodplains exposed. Winds kicked up these surfaces and transported billions of tons of fine-grained sediments to nearby uplands, where a deep mantle of new soil material, called loess, was deposited.[2] Wherever glaciers had been present, the land surface was totally new.

The climatic warming that cleared the land of its ice cover also produced a rapid succession of vegetation changes within two thousand years after the glaciers disappeared. First to emerge was a forest-tundra environment of low, herbaceous shrubs mixed with spruce, *Picea*, forest. As the climate continued to warm, the forest-tundra zone retreated northward. Today it can be observed in the Northwest Territories of Canada, a setting somewhat analogous to the Corn Belt's position 120 centuries ago. The last remnants of this ancient forest can be seen in isolated swamps and bogs as far south as Illinois where stands of tamarack, *Larix laricina*, an extremely winter-hardy deciduous conifer, still grow in undrained depressions.[3] The remnant tamarack bog, sometimes within sight of cornfields on the surrounding upland, is a vivid record of environmental history, just as it reminds us of how close we remain (in space or time) to boreal conditions.

Space substitutes for time in two ways that aid in understanding environmental changes like these. A traverse from a cold environment to a temperate one mirrors the changes that occurred in situ as the climate warmed during the Holocene. A traverse from a high alpine tundra meadow down through successive zones of spruce forest, pine forest, oak forest, and perhaps down to grassland or shrub at the foot of the mountain is the same type of

Fig. 5. Corn and soybean fields cover the Altamont moraine on the Des Moines lobe in Boone County, Iowa. Trees and buildings on the horizon mark a crest of this moraine that was formed about 14,500 years ago in a brief readvance of the ice sheet.

environmental gradient, only compacted into a space of tens of miles—instead of thousands of miles or thousands of years. A perfect substitutability of space for time, and elevation for latitude, would make reconstructing past environments comparatively easy. But the exceptions to perfect sequences, whether in altitude, latitude, or time, are so numerous as to make this type of reasoning hazardous.[4]

There is evidence that the groupings of flora and fauna we recognize as natural environments today may be of fairly recent origin and thus the gradients (ecotones) between contemporary vegetation types may have been different in the past as well. One explanation for this condition is that the full range of climates that have been present during all of earth history have not necessarily occurred simultaneously. Climates existed in Pleistocene North America that have no close, modern counterparts. To the extent that flora and fauna are a function of climate, periods of fairly rapid climatic change may see a rearrangement and regrouping of the species as each seeks to find a new niche that is tolerable. If there is no safe haven for as long as may be necessary to evolve an adaptive response, or if routes of dispersal are cut off, the species is faced with extinction as its surroundings change.

The climate of central North America about twelve thousand years ago was in transition from cooler to warmer average temperatures, but it was also changing in terms of seasonal variation. The hot summers and cold winters of today apparently did not exist in late glacial times. The Great Plains had a cooler and moister summer, yet its winter was not marked by bitter cold. And although the climate may have been rather dismal, with cool, cloudy conditions prevailing, it was mild enough to permit temperate-zone broadleaf deciduous trees to live in the boreal forest. No extensive zone of such an environment is found in North America today. It is plausible that climatic change so altered the conditions for life—and, most importantly, altered them so rapidly—that various species of animals and plants found no place to survive.[5]

Forty-six genera of small mammals and fifty-three genera of large mammals (adult body weight over twenty pounds) have disappeared from North America during the past 3.5 million years.[6] Most of the small mammals disappeared more than 100,000 years ago, whereas the larger ones have extinction dates clustered around 11,000 B.P. All of the continent's *Proboscidea*, including the mastodon, *Mammut*, and mammoth *Mammuthus*, disappeared between 25,000 and 7,000 years ago, with the period of greatest extinction occurring between 11,500 and 9,000 B.P.[7] The pattern of these extinctions—an early and rather gradual disappearance of smaller mammals that continued throughout the Pleistocene, followed by a sudden extinction episode involving most of the larger mammals at the end of the Pleistocene—seems to be correlated with environmental change. But it so happens that the large mammals that are known to have been hunted were those that started disappearing rapidly about 11,000 years ago. Only one new predator, human hunters, entered North America about that time.[8]

Among these late-Pleistocene mammalian extinctions, none has attracted more interest than has the disappearance of the great pachyderms, the mammoth and the mastodon. Sun-bleached bones of these creatures were curiosities to the Indians and the subject of much speculation by the first white explorers. In 1751 land-company scout Christopher Gist carried for a full two months on his return trek to Virginia a four-pound mastodon tooth that had been collected from among the great quantity of bones scattered around the Big Bone Lick in northern Kentucky.[9] Remains of other Pleistocene megafauna suggest geographically extensive ranges for the extinct taxa, although it is difficult to correlate the last known refuges with broader environmental patterns. It might not have occurred to Gist then, but the bison that he observed grazing in great herds would themselves be no more than a memory in the Ohio Valley within another fifty years.

Natural history's time rate of change is faster when humans are the force behind it, although the relative contributions of people versus natural factors of the environment are difficult to segregate when nature itself is changing rapidly. It is a problem that plagues interpretations of the late Pleistocene–early Holocene. Most of the extinctions took place when the climate was changing rapidly, but also at the same time that early humans arrived on the scene. Were the beasts slaughtered? Or did their range simply shrink down to a point and they died?[10] The "overkill hypothesis"—that megafaunal extinctions occurred from hunting—need not take precedence over the evidence on environmental change, because any reduction in the number of animals due to climatic change would only heighten the risk of demise by hunting for those remaining. Faced with a diminishing quantity of their primary resource, hunters would likely develop ever more skillful means of exploiting it. While either condition alone might have produced extinction (and surely some of the species that disappeared after 12,000 B.P. would have been unlikely targets for hunters), together hunting and environmental change were a deadly combination.

The tightest cluster of dates suggesting the antiquity of big-game hunters in the United States is that associated with the Clovis complex identified near that city on the High Plains of New Mexico in the 1930s, following an earlier find near Folsom, New Mexico, in 1926. With the invention of radiocarbon dating methods and much new evidence unearthed by archaeologists, principally in the American West but at a few Eastern sites as well, the Clovis tradition, which is identified by the archaeological presence of distinctive, fluted projectile points used for hunting game, appears to have been widespread in North America around 11,000 B.P. Mammoth or mastodon remains are associated with a number of the sites, to the extent that it seems plausible that many Clovis hunters focused on these vanishing species of elephant.[11] As the Pleistocene megafauna they preyed upon became extinct, so too did the Clovis hunter type disappear. More specialized hunting strategies, peculiar to the new associations of flora and fauna that emerged in the Holocene, replaced the single complex.

Because the dates of Clovis and Folsom remains are nearly all grouped between 11,500 and 10,000 years ago, and because there is no systematic geographical variation in dates for sites ranging from Arizona to Massachusetts that would allow for a gradual diffusion, it must be inferred that Clovis hunters spread not only widely but rapidly. Equally remarkable was the sequence of vegetation changes that followed in the wake of the Clovis people's eastward spread across the central United States during this period of rapid climatic change. The succession that might be expected in a shift to a warmer and drier climate would follow the sequence boreal forest, temperate deciduous forest, prairie grassland. But in the northern Great Plains the deciduous forest phase was skipped entirely, with prairie directly succeeding spruce forest about 11,000 B.P. Fossil pollen reconstructions indicate that a broadleaf deciduous forest of oak, ash, and elm advanced northward from the southern refuge it had taken during Pleistocene full-glacial conditions, and it replaced the boreal forest of northern Iowa, Illinois, and Indiana by 10,500 years ago.

Fig. 6. Bur oaks and other species stand in a relict "oak opening" that has been maintained primarily by grazing, in Dane County, Wisconsin. Open woodlands like these, called *champaigns* by the French, were common in the prairie-forest border zone at the time of European contact.

The prairie, advancing even more rapidly, but from west to east, entered western Minnesota by 10,000 B.P., northern Illinois by 9,000, and appeared at the Ohio-Indiana border 8,000 years ago (fig. 6).[12]

If the prairie had been advancing as a solid front from Minnesota to Ohio, and at a constant rate, that would make an advance of about 150 feet every month for 2,000 years. Of course, the prairie did not advance in such a constant fashion. Its mere presence at a given date, interpreted from the percentage of grasses and prairie forbs in pollen samples collected at widely separated sites, does not suggest that a treeless grassland was present. Although the exact nature of the plant cover is unknown, it is likely that patches of prairie were scattered within an open-canopy deciduous forest, primarily composed of oaks, throughout this period in the Middle West. But a prairie cannot march like an advancing army and wipe out a forest. Grasses need light, trees give shade, and the succession that is universally observed is precisely the opposite: trees invade grasslands.

The explanation of prairie advance that has been favored by most people who have studied this problem rests on evidence of climate, especially climatic change. A shift from a cool, moist climate to one that is warm and dry produces conditions for the desiccation of vascular plants. Prolonged droughts reduce the moisture available for trees to grow. Studies of plant dynamics during the droughts of the 1930s confirmed how brief a period of severe stress is needed to produce vegetation change. A die-back of the forest canopy produced by a succession of droughts would allow grasses to invade.[13] The climatic explanation of prairies reasons that, since grasses dominate the dry Great Plains region today, while forests are found in the humid East, grasses occupy environments that are too dry to sustain trees. Any advance in grasslands should be accompanied by a drier and warmer climate, while an advance of the forest would follow the reverse climatic shift.[14]

Drought in the prairie-Corn Belt of the United States has a specific, known cause. The region's weather is produced by the relative dominance and interaction of three types of air masses.[15] Two of the air masses, in combination, cause precipitation: cold, dry air from the Canadian Arctic hugs the ground and lifts above it warm, moist air from the Gulf of Mexico to produce rain or snow during periods of day-to-day variability in the weather. The third air mass originates over the Pacific Ocean and moves toward the east across the central United States. Pacific air is fairly mild, and it carries abundant water vapor that condenses when clouds are lifted up the west-facing slopes of the Sierra Nevada and the Rockies. But once Pacific air descends the east-facing Rocky Mountain front, the last topographic barrier, its temperature warms again and it is unlikely to produce rainfall. When the Pacific air stream dominates the mid-continent, day after day of dry conditions are the norm, and the pattern can last long enough to produce serious drought.

This three-air-mass regional weather system did not exist during the Pleistocene, when mountain-range-sized ice sheets blocked the flow of Arctic air into the mid-continent. As the ice sheet wasted northward around Hudson Bay, westerly flows dominated by the Pacific air stream began to control middle-western weather most of the year. The period from about ten thousand to seven thousand years ago saw increasingly frequent droughts as a result of dominance by the westerlies. That interval coincides with the advance of the prairie from the Great Plains to Ohio. Modern atmospheric circulation patterns over eastern North America began to emerge after 6,000 B.P.: warm, moist air from the Gulf; cold, dry air from the Canadian Arctic; and mild, dry air from the Pacific.[16] Droughts became uncommon on the northeastern and southeastern margins of the Mississippi Valley, where the drying effects of Pacific air were ameliorated by the Arctic and the Gulf air masses.

The area of most frequent Pacific air dominance took on the shape of a wedge covering the entire Great Plains, which narrowed progressively eastward to an apex near the southern edge of Lake Michigan. The significance of the Pacific air wedge is that its outline matches closely the wedge-shaped outline of the North American prairie. Grasses dominate all of the smooth land of the western Great Plains, but the latitudinal breadth of the grassland narrows toward the east, culminating in the Prairie Peninsula of eastern Illinois and western Indiana, an area that is bordered by forests to the north, east, and south. The match between the drought-prone Pacific air wedge and the outline of the North American prairie was first reported by John R. Borchert in 1948, some years before the evidence on Holocene climatic change was accumulated.[17] Studies done following Borchert have confirmed the processes that he suggested and have offered a plausible scenario of the ten thousand years of environmental change that led to the present pattern.

While the association of Pacific air-mass dominance with droughts is well established, the question remains as to whether such conditions can make prairies. Moisture-demanding trees might have perished from prolonged drought; but they might also have been replaced not by grasses but rather by smaller trees and shrubs that would have invaded from the Southwest as the climate of the Middle West changed. Some deciduous forest varieties evolved adaptations to survive long periods of drought and hence would have been selected under stress. One tree that evolved such an adaptation at some point in its evolutionary history is the bur oak, *Quercus macrocarpa*, a tough competitor in supposedly treeless environments of the Great Plains; it is found throughout the mid-continent prairie zone today. The bur oak has a deep tap root that extends down far enough to reach the zone of moist soil— an obvious advantage in a drought-prone environment—and it also has a thick bark that is protective enough to enable it to survive searing fires that would destroy many other woody varieties.[18] The bur oak can survive both fire and drought (fig. 6).

Of these two natural forces, fire is the more effective in totally eliminating woody plants from the scene. Moreover, grasses sprout readily on newly burned lands. Repeated burning increases many grasses and forbs just as it eliminates woody competitors. The equilibrium is stable in the long run with repeated applications of fire. A treeless grassland is established. The equilibrium is destroyed when regular burning ceases: woody plants invade and the forest eventually is reestablished. Grasses can exist up until—but not beyond— the time woody plants form a continuous canopy of shade. A treeless grassland of the sort the early white explorers reported in many areas of North America would need fire at frequent intervals in order to be maintained. Numerous mentions of Indian-set fires in the journals of early European and American observers leave no doubt of the frequency of fire and its role in maintaining the prairies after the middle of the sixteenth century A.D.[19]

But there is little archaeological evidence linking fire with the creation of the prairie patches that emerged in the northern Great Plains by eleven thousand years ago, the time at which Clovis hunters were abroad in the land. Nonetheless, there are reasons beyond the coincidence in timing to examine the possibility of human intervention in the creation and extension of the prairies. More than fifty years ago Carl Sauer advanced his thesis that the grasslands were the product of early man's practice of firing to drive game. Sauer was neither the first nor the last to make such a suggestion, but the "cultural grasslands" thesis has had few followers.[20] It has played almost no role in the research of the past several decades that has sought to link vegetation change with climatic change.

One reason for taking Sauer's argument seriously is that fire would have been especially effective in driving herds of large, gregarious mammals such as the bison, and perhaps even more so against still larger, ungainly creatures like the mammoth. Herds stampeding downwind with the lick of fire at their heels would be weakened and more easily slaughtered with the weapons at hand. Fire would be no great assist in hunting either small game or animals unaccustomed to massing in herds. But it would be of double benefit if the prey were the large herbivores of herd instinct like the bison: fire assisted in the hunt and it also increased the nutritional value of the grassland in the next season by eliminating the dead plant remains.

Of all the occupations known to have been practiced by aboriginal North Americans during the past ten thousand years, none is more associated with the prairies than is bison hunting. One or more species of bison appear to have ranged from Canada to Florida and from Pennsylvania to California during the late Pleistocene. Bison remains have been identified at a number of Clovis sites. The many mentions of bison in the journals of early explorers establish an enormous range for the animals, from the Appalachians to the Rockies and from Mexico to Alberta.[21] Evidence suggests that this range was not constant nor was it at all times coextensive with the North American grassland we know today. During some prehistoric periods bison hunting was less evident on the Great Plains than it was in the eighteenth and early nineteenth centuries. There are lapses in the record of big-game hunting at specific localities, perhaps the result of environmental fluctuations or the roving patterns of small numbers of hunters. Even during the time between visits of sixteenth and seventeenth century European observers there were changes in the extent and quality of the range. New grasslands supporting bison herds appeared where 150 years earlier only a valueless scrub woodland had been recorded.[22]

The various retreats and advances of bison habitat were much the same as the retreat and advance of the grasslands, which, in turn, may have been produced by the human activity of setting fire to the range. Natural (lightning-caused) fires occur in grasslands, although their frequency cannot be termed common. Both natural and human-caused fires would have been favored by the drying and warming episodes of Holocene climate. The best conditions for the rapid spread of fire are a smooth land surface, continuous vegetation cover, consistent wind direction, and an abundance of dry combustibles. The interior plains of the United States met all of these conditions as soon as the rapid transition to the Holocene began. Fire, whether culturally or naturally caused, appears to be a necessary part of the explanation of the grassland's existence. No other force, including droughts, could have totally eliminated woody plants from the scene. Since a smooth land surface is just as important as climate in explaining why grasslands are created by fire, the association between prairies and smooth plains in the Middle West acquires greater significance.

If aboriginal hunters had the incentive to use fire and occupied an environment where fire would spread, they also possessed a strategy that worked better the more it was used. The early post-Pleistocene plant cover consisting of trees, shrubs, and grasses was a continuous one through which a fire would have spread rapidly to the east, given the prevailing wind direction. Burning eliminated most of the woody perennials and increased the grasses. Repeating the procedure could only have improved the range for grazing animals, so their numbers are likely to have increased. A bison overkill would have been extremely unlikely under the conditions of an expanding grassland. Increasingly throughout the Holocene, big-game hunters became masters of the gently rolling uplands, the kind of topography over which fire spreads easily. Steep valley sides stopped the fire's advance and protected the bottomlands along the rivers, although deliberate clearing and burning of flat bottoms eventually produced grasslands there as well.

Sauer's belief that the grasslands were advanced by humans was as controversial as his tendency to push back in time their entry into North America. While belief in a much greater antiquity of occupancy remains unproven, later arrival makes it unnecessary to explain the long hiatus between entry and the emergence of grasslands. Nor does the recent evidence of only a three-thousand-year lag between prairie patches in the northern Great Plains and similar clearings east of Lake Michigan make it any less necessary to postulate some triggering mechanism that would transform an evolving deciduous forest into a degraded woodland in which prairie plants were gaining ascendancy.

Fires were favored by the Holocene dry conditions. They shredded the forest margins and enabled grasses to spread east across the flatlands wherever the hardwood forest canopy was opened. The advance was rapid at first, but it slowed in Indiana and came to a halt in western Ohio. Neither state ever had as much prairie as Illinois. The prairie-forest border seems unlikely to have been a definite line, but rather a patchwork of occasional treeless plains, hardwood groves, and tracts of savanna (open woodland) nearly everywhere fringed by strands of forest that encroached upon the flat uplands from pro-

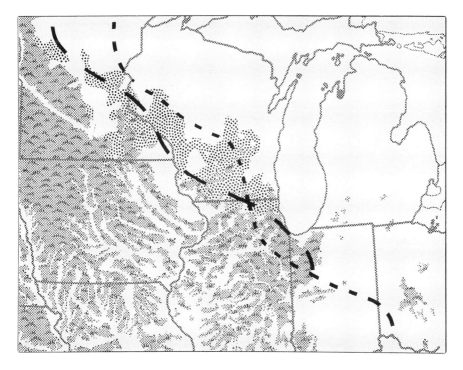

Fig. 7. Ten percent isopolls for prairie forbs 7,000 B.P. (dotted line) and 3,000 B.P. (dashed line) superimposed on map of historic prairie and savanna (shown in dotted pattern) vegetation in the Middle West. The wedge shape of the Prairie Peninsula has emerged in the past 3,000 years. (Prairie and savanna modified from Kuchler, *Potential Natural Vegetation*, 1969; isopolls based on Webb et al. [see chap. 2, note 12].)

tected valley sides whenever burning ceased. The solid stands of timber on the floodplains responded to climatic fluctuations in terms of changing fluvial regimes but were unaffected by upland fire where fringed by valley sides steep enough to prevent a fire's downhill encroachment.

These post-Pleistocene environmental changes were not once and for all. The eastern edge of the prairie has fluctuated during the Holocene (fig. 7). For example, Indiana had a significant amount of prairie seven thousand years ago, but it was invaded by forest later as the grassland retreated to the west until the present Prairie Peninsula outline was reached.[23] The mid- to late-Holocene shift toward a more pronounced domination of Gulf and Arctic air masses over the eastern United States, with a correspondingly decreased role for Pacific air, produced wetter conditions, favoring true forests in northern Indiana. The extra precipitation, combined with less frequent drought, decreased the effectiveness of fire. In this instance, again, climatic shifts would seem to permit vegetation changes, although they alone do not account for them. The farther to the east the grassy bison range expanded, the more certainly one can point to cultural, rather than natural, factors as the cause.

The expansion of grassy openings within the eastern woodlands is not the only modification of the plant cover to be found. The bison themselves were agents of change. Even the migratory habits of the enormous herds left their mark on the land as thousands of hooves trampled everything underfoot, created trails through difficult mountain gaps, forded streams, and scarified the land around the salt licks. They set a straight-line course wherever it was possible, discovered the easiest passes and fords, and generally left an unmistakable network of good trails (traces) that first the Indians and then the Europeans traveled in pursuit.

Wherever the bison were, there too were found modifications in the local biota. In Marquette's relation of his voyage with Joliet down the Mississippi in 1673, he wrote of the "canes or large reeds, which grow on the bank of the river . . . and grow so thickly that the wild cattle [bison] have some difficulty in forcing their way through them."[24] Cane, *Arundinaria* spp, probably occurred naturally in the alluvial floodplain of the lower Mississippi, below its confluence with the Ohio, but it spread farther inland along streams that entered the lower Ohio and Mississippi. Canebrakes appeared up the valleys followed by the migrating bison as far as the salt licks of Kentucky and middle Tennessee. The bison were effective disseminators of plants that took hold wherever the animals, through their sheer numbers and habit, provided a fresh surface.

Bluegrass, *Poa pratensis*, was not mentioned by the astute John Filson in his 1784 description of Kentucky (bluegrass, of European origin, came later, along with white settlement from the East). Rather, Filson wrote of the "great plenty of fine cane, on which the cattle feed, and grow fat."[25] He described the plant as growing three to twelve feet high, composed of a hard substance, with joints along the stalk "from which proceed leaves resembling those of willow." Notations of cane land appear on his map in several places, from the upper forks of the Licking River, to the Inner Bluegrass near present-day Lexington, to the forks of the Salt River near Bardstown—localities spanning the entire breadth of what is now called the Bluegrass section of Kentucky.

A similar scene was to be found in the Nashville Basin of Tennessee. The first white settlement was a trader's fort at the French Lick, a sulfurous spring near the Cumberland River in what is now downtown Nashville. One of the Bledsoe brothers, members of the group known as the Long Hunters, who roamed the low plateaus of Kentucky and Tennessee in the latter half of the eighteenth century, visited the French Lick. Bledsoe found "an area of about one hundred acres so crowded with buffalo and other animals that he was afraid to dismount from his horse lest he should be run over and trampled to death."[26] The lick was named with reference to the French traders from Illinois who came there to slaughter the bison. They used just the tongues and tallow and left the discard to rot. The bleached skulls and bones that lay about the fort had accumulated to such a thickness that Bledsoe could walk upon them for a distance of several hundred yards.

Canebrakes surrounded the lick and the bone heaps for miles up and down the banks of the Cumberland. The first stations (fortified outposts) built by the several hundred settlers who came there in 1779–1780 were in cane so tall that the stations could not be seen one from another. Farther out there were clover bottomlands and broad rolling uplands covered in a luxurious carpet of grass. Cedar glades rooted in thin soils on the fringing bluffs must have contributed to the park-like appearance.[27] Nearly all the attractions the Nashville Basin held for its early settlers resulted not from nature alone but from how the place had been used—by animals and by those who hunted them. It surely was no primeval landscape.

Salt springs and the saline streams that trickled from them were found widely in the limestones and shales of the Ohio and Mississippi valleys. Between the salt licks there were low limestone plateaus, broken in places by the valleys of deeply entrenched streams; the plateaus offered broad expanses across which fires burned rapidly and bison moved just as fast. The favorite ground of the Long Hunters was the Pennyroyal Plateau, the low upland that marks the drainage divide between the Green and Cumberland rivers. The annotation on Filson's map reads, "Here is an extensive Tract call'd Green River Plains, which produces no Timber, and but little Water; mostly Fertile, and cover'd with excellent Grass and Herbage." It was called the "Kentucky Prairie" by the forty Long Hunters who roamed it in 1770. Thousand-head herds of bison and an abundance of elk, white-tailed deer, and turkeys were targets for the adventuring hunters. Skulls and horns of bison and elk along with charred tree stumps were reminders of those who had been there before. In his text, Filson refers to the area as the "Green River Barrens," a label that might suggest an impoverished flora, although that word had no such connotation in Virginia, where "barrens" was used to denote a variety of lands without timber (fig. 8). The Long Hunters, and their successors, rendered the place barren in another sense: by 1793 not one bison remained alive.[28]

The Pennyroyal/Barrens is significant for several reasons. An outlier of the prairies to the northwest, it was surrounded by old-growth hardwoods when it was first recorded. Nothing about the place suggests that its prairie was the product of climate or of climatic change. It was the first extensive tract of grassland settled by the white population of North America. Nathan-

Fig. 8. The forest has long since been reestablished in Barren County, Kentucky, a part of the Pennyroyal Plateau along the southern edge of the state. In 1784 John Filson called it the "Green River Barrens," a vast, open prairie at that time.

iel Shaler recognized it as the product of deliberate burning and added that in the years since settlement came and burning ceased the forest had nearly regained its former ground. It was in the Pennyroyal that Carl Sauer undoubtedly formulated his theory of grassland origin.[29] Far from being repulsive for being barren of large trees, it was the first choice of Virginia for a Revolutionary War military tract west of the Appalachians. By 1820 one-fourth of the population of Kentucky lived there.

John Filson reserved his greatest praise for the section of Kentucky he knew best, the Outer Bluegrass: "The reader, by casting his eye upon the map, and viewing round the heads of Licking, from the Ohio, and round the heads of Kentucke, Dick's River, and down Green River to the Ohio, may view . . . the most extraordinary country that the sun enlightens with his celestial beams."[30] The condition of the land, apart from the abundance of cane, bison, and salt springs, has had conflicting descriptions, however. None of the early accounts suggests that the Bluegrass was "barren" like the Pennyroyal. Gilbert Imlay described the northern Bluegrass as "immensely rich, and covered with cane, rye grass, and the native clover." The Inner Bluegrass around Lexington was to Imlay "the finest and most luxuriant country, perhaps on earth." Timothy Flint thought "the woods [had] a charming aspect, as though they were trees, promiscuously arranged for the effect of a pleasure ground." But in his annotations on Filson's work, geologist Willard Rouse Jillson suggested a very different landscape, one that was "densely forested with giant broadleaf trees."[31] If so, the area was unique in having large numbers of bison without also encompassing substantial open spaces.

The Nashville Basin, Pennyroyal, and Bluegrass had a similar pattern of human occupancy well back into prehistoric times, and all three have histories involving occupancy by the Shawnee dating to the mid-seventeenth century. There was a Shawnee village at the French Lick on the Cumberland as early as 1665. The Cumberland River itself was known as the Shawnee (Riviere des Chaouanons) to the French, while the Kentucky River was identified as the Cuttawa on Pownall's 1776 map. Both "Kentucky" and "Cuttawa" appear to be descriptive of the landscape. Darlington states that Kentucke is a Mohawk word signifying "among the meadows," while "M'shish'-kee-we-kut-uk-ah" was identified by Schoolcraft as a Shawnee designation for meadows.[32]

Where were these meadows along the Kentucky River that would suggest the name? The river itself flows in a deep, meandering gorge trenched down across the axis of the Jessamine Dome, the geologic uplift on which the Bluegrass rests. Imlay's description rings true: "the Kentucky is bound everywhere by high, rock precipices. . . . Few places on it have any bottom land, as the rock rises most contiguous to the bed of the river."[33] Whatever meadows there were had to have been on the gently rolling uplands, likely in the region now known as the Bluegrass. And if the linguistic evidence is to be believed, the meadows were there a long time before the first Europeans arrived, probably even before the Shawnee, although the local environment in no way suggests that the meadows were of natural origin.

The Shawnee had lived in the upper Ohio Valley but were driven out by the Iroquois. Some fled to the west, along with other tribes so displaced, while the Iroquois compelled some other Shawnees to move east, principally to Pennsylvania. The geographical range of Shawnee impact was extensive. "Scioto" appears earlier as "Ona-Sciota" or "Ouasiota," an erstwhile name for the section of Appalachian Kentucky now known as the Cumberland Plateau.[34] Around 1730 their forced residence in the East ended and, as the Shawnee began to leave Pennsylvania, they reestablished themselves in the Ohio Valley they knew so well, making a new village at the mouth of the Scioto River near present-day Portsmouth. From there the Shawnee dispersed northward after 1758 (perhaps joined by kin who had fled to Illinois) and built several villages between the Great Miami and Scioto rivers. Although they hunted in Ohio, the Shawnee also resumed their old habit of bison hunting in the Bluegrass. It was during this final stage of reestablished Shawnee occupation that Daniel Boone and company arrived. What had obviously been a hunting paradise for the Shawnee acquired the epithet "dark and bloody ground."

At one time or another the Shawnee were found in all five of the areas that would first attract Euro-American agricultural settlers west of the Appalachian Mountains: the Nashville Basin, the Pennyroyal Plateau, the Bluegrass, the Scioto Valley, and the Miami Valley. The maize agriculture that the white settlers would bring with them owed much to the Indians in general, although little to the Shawnee in particular. Rather, what the Shawnees—and their unknown predecessors who hunted the same ground even earlier—gave to their white usurpers was an entire landscape, substantially modified by the uses to which it had been put. These Indians, like so many others, left to a civilization that would ultimately destroy them a country much improved over what nature alone had provided, one that was ready for the plow immediately. No wilderness that demanded hard labor to clear, these five areas looked nothing like the rugged Appalachians, nothing like any but a few of the long-settled areas of the eastern seaboard. The expanses of rolling grassland left no doubt what should be done there, at least not in the minds of men who envisioned farming on a grand scale. It was in these five islands of plenty that the Corn Belt was born.

3

Finding the Land

THE FIRST VIRGINIANS accepted on faith a premise they shared with all other Europeans who sailed the Atlantic Ocean. They supposed that the North American land mass was very much smaller than it is and that the shores of a southern ocean lay not too far back into the interior. Powhatan knew better, and well before the end of the seventeenth century so did the Virginians. But the 1609 charter that the London Company had obtained from King James I extended the western limit of Virginia to the "South Sea." Later this was interpreted to mean that "the charters of . . . Virginia never had any serious western limit, since the South Sea was thought to be nearer the Atlantick than it really is; and if its true position had been known, such a grant would have been too extravagant."[1] The issue of extravagance would enter nearly every controversy over Virginia's western land claims that would emerge once the westward movement got under way. The land claims issue would not be resolved until the bounds of the Virginia Military District in Ohio were fixed in 1784, more than a century after serious explorations began.

When Dr. Thomas Walker "discovered" Cumberland Gap on April 13, 1750, he crested the grade and walked a short way down the western slope to cut his name into a beech tree. He may have intended to leave his mark on a tree higher up, closer to the pass, but at the top he found "Laurel Trees marked with crosses, others Blazed and several Figures on them," signs of the woodsmen who had been there before him. The first trans-Appalachian expedition of record was that of Captain Abraham Wood, which crossed the Blue Ridge and found the New River in 1671. Wood's party reported numerous beech trees carved with initials.[2] Regardless of who made the discovery, it was clear well before Walker's time that an uncharted wilderness, not the South Sea, lay at Virginia's back door.

The first step toward geopolitical reality came when the assumed South Sea was replaced by knowledge of a large land mass. Desire to possess the land followed immediately the first awareness of its existence, although little thought was given to the nature or quality of this great interior domain until early in the eighteenth century. A sylvan wilderness populated by Indians suggested the usual possibilities for trade, but in a short time even this sort of reciprocal linkage between territories would not be enough. To possess the land meant to bring it under cultivation, and before that could take place it was necessary to know where the best lands would be found so that the new society on the western frontier would have a strong base upon which to prosper.

Each new piece of intelligence gathered from the interior dispelled old myths, but also set new expectations for what might be possible. The first solid information came by accident, in the reports of parties searching for an ocean but finding only mountains and rivers. As an awareness of the interior took shape, expectations about its nature rose in proportion. Western explorations became more deliberate. By the middle of the eighteenth century it was no longer enough to note the general locations of rivers, mountains, or favorable passes. Detailed explorations were needed to search for specific sites where capital investments could commence swiftly and profitably.

By the time the first English colonials ventured west to the Mississippi River's drainage the French (not to mention the Spanish) already had a good idea of what the country was like. One of the reasons the court of King George II was favorably disposed toward making grants of land in the interior was the knowledge of French penetration. Virginia and Pennsylvania lay closest to the western interior of any of the seaboard colonies. While Pennsylvanians were busily engaged in trading with the Indians in the Upper Ohio Valley, Virginians began to make plans not only for trade but also for planting colonies, similar to those along the seaboard but located west of the Atlantic watershed. Two land companies were formed to petition the king for tracts of western land; both were successful, and on July 12, 1749, the governor and council of Virginia made the grants.[3] The Loyal Company, formed by a group of politically prominent southern Virginians, received a grant of 800,000 acres in southwestern Virginia, including what is now part of Kentucky. The Ohio Company, comprised of men having equal political skill and comparable ambitions for land, but who lived in the Northern Neck of Virginia (north of the Rappahannock) and Maryland, received 200,000 acres in the Ohio Valley.

Both companies were organized land speculations that involved the Virginia gentry.[4] Among the twenty-five Virginians who held stock in the Ohio Company, twenty were at one time or another members of the House of Burgesses. Ohio Company members included Thomas Lee, who induced migration into the northern Shenandoah Valley; Lawrence and Augustine Washington (half-brothers of George); Thomas Cresap, a Maryland planter who held substantial acreage along the Susquehanna; George Fairfax, heir to a large grant in the Northern Neck; and George Mason. The Loyal Company's members included Peter Jefferson (father of Thomas), Edmund Pendleton, John Lewis, and Dr. Thomas Walker. The two groups were political rivals who reflected different sectional interests in Virginia just as their western land ambitions were projected in different directions. But what lay beyond the Atlantic watershed in the domains these men coveted was largely unknown to them in 1749.

Even before the grants were approved, both companies planned expeditions to explore the country. The Ohio Company's first effort, in 1749, was aborted. In the spring of 1750, however, Thomas Walker embarked on a four-month reconnaissance for his Loyal Company. Evidently trained as a physician, but also known for his skills as a surveyor, Walker lacked imagination but had a good eye for his surroundings. He and his party followed the traces and Indian trails through the Ridge and Valley section, crossed through Cumberland Gap, and entered the drainage of the Cumberland River in eastern Kentucky. His journal is faithful to details of the countryside, although he registers scant enthusiasm for what he saw.

April 19, 1750: "In the Fork of Licking Creek is a Lick much used by Buffaloes and many large Roads lead to it. This afternoon Ambrose Powell was bit by a Bear in his knee."[5] In the vicinity of Barbourville, Kentucky, the party built a cabin and planted corn and peach stones, thus evidencing a claim to land of the sort recognized in Virginia. Later, two of Walker's horses were bitten by snakes, one on the nose; a bear broke one of his dog's forelegs. Rainy weather, rough land, and swollen streams that had to be followed for miles before fording seem to have impressed Walker the most. He makes two references to "burnt woods," noting a regrowth forest on an old burn in one instance, and on May 30: "The Woods are burnt fresh about here and are the only fresh burnt Woods we have seen these Six Weeks."[6] Walker had to have been familiar with the practice of burning the woods for driving game, as was the habit of Indians in Virginia. An evidence of burning thus was a sign of human presence. Although Walker did not make the statement, it would seem that he interpreted a lack of burning as a sign of little human activity. That, in turn, would not be a favorable omen to a land-company scout who was seeking territory that would offer good support for people who would live there later.

No one knows for sure exactly where Walker's wanderings led him in Kentucky. It has been suggested that he saw the Bluegrass section, but that seems unlikely since he makes no mention of the kind of landscape that those who actually went there recorded later. No clover meadows, canebrakes, or even the suggestion of broad clearings appear in Walker's journal. Instead, he makes repeated references to impenetrable laurel slicks and ivy covering steep slopes. He crossed many streams that would have taken him down to the Bluegrass, but more than likely he made a great circle around their upper reaches and returned to Virginia never having gotten beyond the eastern edge of the Interior Low Plateau. If Thomas Walker was impressed with Kentucky, he never entered any thoughts to that effect in his journal. His was a trek through the wilderness.

The Ohio Company's trans-Allegheny expedition was begun in September 1750, under the leadership of Christopher Gist, who, like Walker, was known as a surveyor of considerable skill. Gist was a Marylander who had received a quantity of land at the time of his marriage, but after suffering business reverses he had moved south to the Yadkin River country of North Carolina. From there he was summoned by the Ohio Company and given some specific instructions by Governor Robert Dinwiddie as to what his journey should include:[7]

> When you find a large quantity of good, level Land, such as you think will suit the Company, You are to measure the Breadth of it, in three or four different Places, & take the Courses of the River & Mountains on which it binds. . . . [T]he nearer in the Land lies the better, provided it be good & level, but we had rather go quite down the Mississippi than take mean broken Land.

Land—undifferentiated as to quality—was no longer the objective. It had to be good land, fairly level in slope, and in bodies as large as possible. The acceptable trade-off between land quality and remoteness is also instructive: the Ohio Company's leaders were more interested in good land than they were in accessible land. Such criteria would not have been of interest to a company planning only to build trading posts or to extract mineral resources—or even to immediately plant an agricultural colony as close to settled territory as possible. Dinwiddie's instructions were appropriate to men whose interests were in the long term as much as in the short, and who envisioned large-scale agriculture on good lands rather than scattered cabins in the clearings of a dense forest. The criteria stand in sharp contrast to those of the trade-oriented French, who had equipped an expedition south from Montreal in 1749, led by Captain J. P. Céloron de Blainville. Céloron's instructions were to bury inscribed leaden plates at the mouths of all rivers tributary to the Ohio, thus claiming the land for France ahead of the English.[8] A claim of the sort France was making was relevant to traders who sought a business monopoly, but it meant little to men who planned permanent agricultural settlement.

Dinwiddie and his associates in the Ohio Company knew what the results of Walker's trip had been, and they hoped that Gist would find a country that was more attractive. Dinwiddie added that Gist was to "note all the Bodies of good Land," even though the company's grant would not be large enough to encompass all the desirable land in the Ohio Valley. Such ambitions recall the comment made in 1705 by Virginia's historian, Robert Beverley, who wrote of his fellow citizens that they did not mind anything "but to be masters of great tracts of land."[9] Beverley's point was to condemn the dispersed pattern of Virginia's plantations, wherein "the ambition each man had of being lord of a vast, though unimproved territory" had left the Virginians without "one place of cohabitation among them . . . that may reasonably bear the name of a town." The Ohio Company's interest in large bodies of good, level land would be expected from a group of men who already held the best lands in Virginia and wanted more. And, as in Virginia, it was tracts of good land, rather than appropriate sites for towns, that attracted the attention of investors.

Gist was instructed to explore "the Lands upon the river Ohio," with no restriction as to which side of the river he should take, other than the stipulation that he go downstream as far as the falls (Louisville). He began his journey in October 1750, crossing from the Potomac to the Monongahela watershed, all the way making observations on the quality of the land. He noted the landscape fairly, as in his description of the Forks of the Ohio, the site of Pittsburgh: "Bottoms not broad; At a Distance from the River good Land for Farming . . . and tolerable level."[10] Perhaps because he had been instructed to meet the Indian trader George Croghan, who was in Ohio at the time, Gist left the Ohio River near the Indian village of Logstown, at the river's northernmost bend, and headed straight west into Ohio.

After joining Croghan, the party visited several Indian villages. Gist pretended his trip was diplomatic, to spread good will, not one of detailed survey intended to spy out tracts of land. He kept his compass hidden from the Indians, knowing that they understood its purpose. Gist was following by little more than a year the Céloron expedition that had the purpose of establishing French authority and warning away any English traders they encountered in the Ohio country. Céloron had told the native inhabitants that the English would come and take their lands. After Gist made ranks with Croghan (who seems to have been sent by the Governor of Pennsylvania to disabuse the Indians of what Céloron had quite accurately predicted about English intentions the year before), the party continued west across Ohio. Their course was fixed by the need to visit the Indian villages, but Gist kept detailed descriptions of lands encountered on the journey, nonetheless. He produced a remarkable journal that was faithful to the charge he had been given.

On January 20, 1751, the party reached the Delaware's town of Maguck, on a level plain above the east bank of the Scioto River just south of where Circleville, Ohio, stands today. After having had little to praise in the hills and woodlands of eastern Ohio, Gist was amazed to see next to the Delaware's village "a plain or clear Field about 5M in Length . . . and 2 M broad" along the Scioto River. This plain, which occupies two levels of alluvial terrace on the east bank of the Scioto, was known variously as the Shawnee Plains (they had no villages there at the time of Gist's visit), the Great Plains of Maykujay, or, after white settlers arrived, the Pickaway Plains.[11] Without

Fig. 9. The Pickaway Plains along the Scioto River south of Circleville, Ohio. Christopher Gist wrote favorably of this place on January 20, 1751, the first Anglo-American discovery of good land west of the Appalachians.

intent of doing so, since his destination was the Indian settlement and not the plains, Gist had discovered one of the most desirable tracts of land in the Ohio Valley (fig. 9). He saw the land in a partial snow cover in January and thus could not describe much about the vegetation, but he instantly recognized the land for its value.

After traveling down to the mouth of the Scioto to visit the Shawnee's principal town, Gist set out toward the northwest on February 12, spending five days traveling toward what is now Logan County, Ohio:[12]

> All the Way . . . to this Place (except the first 20 M which is broken) is fine, rich level Land, well timbered with large Walnut, Ash, Sugar Trees, Cherry Trees, &c, . . . well watered . . . and full of beautiful natural Meadows, covered with wild Rye, blue grass and Clover, and abounds with Turkeys, Deer, Elks and most sorts of Game particularly Buffaloes. . . . In short it wants nothing but Cultivation to make it a most delightfull Country.

The broken land was the unglaciated upland; the "fine, rich level Land" was the Wisconsinan glacial-drift plains west of the Scioto River.[13] Gist was observing the best agricultural land that any American had yet described.

Gist and Croghan then spent several days at the sizable town of the Twigtwee's (a tribe of Miamis), near where Piqua, Ohio, now stands. After several days of discussions the party headed back south, thus making a second traverse of the same area but staying a little farther to the west on the return. Once again Gist was struck by the landscape that he saw, "very rich, level and well timbered, some of the finest Meadows that can be," and remarked that he had been told the land was just "as good and if possible better, to the Westward as far as the Obache" (Wabash River).

On March 3, after they crossed the Mad River, George Croghan and the rest of the party, having completed their business with the Indians, headed back east to the Forks of the Ohio, leaving Gist to return alone to the mouth of the Scioto, where he had left his African-American servant. Gist had "fine traveling thro rich Land and beautiful Meadows," in which he saw "forty or fifty Buffaloes feed at once." The Little Miami ran through "a fine Meadow, about a Mile wide very Clear like an old Field, and not a Bush in it, I coud see the Buffaloes in it above two Miles off." Bonnécamp's relation of the Céloron expedition of 1749, which also passed through the Miami Valley, recounts a march "over vast prairies where the herbage was sometimes of extraordinary height." Bonnécamp's prairie obviously was Gist's meadow.[14]

After reaching the mouth of the Scioto, Gist continued his journey toward the Falls of the Ohio. He didn't reach there, and he may not even have gotten close. After a rendezvous with some traders in the employ of the Ohio Company (who gave him the mastodon tooth that he carried back to Virginia), Gist headed south and east and soon passed into the hilly lands of the Appalachian Plateau, which pleased him even less than they had Walker. According to Gist, the country was "rocky mountainous & full of Laurel Thickets, the worst travelling I ever saw." A more famous Virginian, George Rogers Clark, who settled at the mouth of the Kanawha River in 1772, was similarly repulsed by the sight of high mountains and rugged terrain.[15] After good lands in the Ohio Valley had been confirmed, the high country of the Appalachian Plateau would be bypassed by the mainstream, left to those content with hunting in the woods or scratching a hoe into the rocky soil on a cleared patch of hillside.

Gist's journal seems to have been read by various members of the Ohio Company immediately after his return to Virginia. The value judgments he made as to the richness of the land, associating mature stands of timber and tall-grass meadows with fertility, certainly would have seemed authentic to those who read his words, then or now. His account was by no means forgotten, but within six months he was preparing for another journey to the Ohio Valley, this time to stay south of the river and "to keep a more detailed description, . . . noting not only every fragment of good land, no matter how small, but also the bad land, describing both as to length, breadth, value, and produce such as timber or trees."[16] While a description of good land would seem to be more useful to land speculators than an account of bad, Gist probably read no criticism of his account of the first journey in the request to make a second one so soon. The real purpose of the second trip was to stake out a definite piece of ground that the company could occupy at once for the purpose of confirming a strategic presence.

At this point there followed an increasingly complicated series of moves by the Ohio Company to assert its presence in the upper Ohio Valley. The French and Indian War and the drawing of the Proclamation Line of 1763, which forbade settlement west of the Atlantic watershed, postponed any efforts to plant colonies north of the Ohio River. The Pickaway Plains saw a different kind of expedition from Virginia in October 1774, when John Murray, Earl of Dunmore, camped there with his militia during the so-called "Dunmore's War" against the Indians. The Quebec Act of 1774 attached the country north of the Ohio to Canada, thus proscribing the western land claims of Virginia or any other seaboard colony.[17]

Shortly after Gist's first reconnaissance, the Shawnees he had visited at the mouth of the Scioto began to move north and west. They built several new villages, or "chillicothes" (chaw-law-Kaw-tha) between the Scioto and Miami rivers. Bison, deer, and other game were found in this territory, but the Shawnees also made seasonal trips to their old hunting grounds in the Bluegrass of Kentucky. Now, however, the land they hunted was no longer theirs unchallenged. White incursion may have been slowed by the British proclamation against settlement, but it was by no means stopped. Virginians established Harrodsburg and Boonesborough in 1775, and small settlements or stations of others soon followed.

The series of battles between Kentuckians and Shawnees over the dark and bloody ground that the Bluegrass had become culminated in a massive expedition led by George Rogers Clark in November 1782. With some 1,050 mounted troops and Daniel Boone, the expedition crossed the Ohio and headed for the Shawnee towns. They captured Old Chillicothe and, according to Filson's life of Boone, they continued their "pursuit through five towns on the Miami rivers, Old Chelicothe, Pecaway, New Chelicothe, Will's Towns, and Chelicothe, burnt them all to ashes, entirely destroyed their corn, and other fruits, and every where spread a scene of defoliation in the country."[18]

Boone knew well this country he was despoiling, partly because of events that transpired after his capture by the Shawnees at the Blue Licks of Kentucky in 1778. The Shawnees took him to their chillicothe on the Little Miami. Later, Boone and his captors traveled and hunted together, apparently in good spirits, across the prairies to the salt springs on the Scioto River. Boone "found the land, for a great extent about this river, to exceed the soil of Kentucky, if possible, and remarkably well watered." Filson had not seen the Scioto Valley but, probably on Boone's word, he entered the notation "Natural Meadow" on his map at the location of the Pickaway Plains.[19]

The vigorous prosecution of Virginia's "rights" in the Ohio Valley under Lord Dunmore and, later, George Rogers Clark, never established a claim sufficient to counterbalance the western ambitions held by the other colonies. Resolution of the conflicting interests of rival groups of land speculators, allied variously with Virginia, Maryland, Pennsylvania, New Jersey, and New York, was a task for the Continental Congress. The practice of projecting territorial claims in straight latitudinal swaths into the interior had been followed by several seaboard colonies and was supported in some cases by dubious treaties with the Indians, who were forced to admit to the authority of some colonial governor. Virginia held to the southern limit of its early charters, which marked the boundary with North Carolina, but Virginia also continued to claim, under authority of charters or subsequent Indian treaties, all of the Ohio Valley, including what is now western Pennsylvania. The issue of western land cessions by the original colonies was taken up in Congress early in 1780, and the point of greatest controversy was Virginia's western lands.

In September 1780, a committee issued a report urging that all of the western land claims be relinquished, but it recommended retention of the so-called military bounty lands of the colonies, lands awarded in the form of warrants entitling Revolutionary War soldiers to specified amounts of unclaimed land in the West. Virginia had set aside the area between the Green and Cumberland rivers (the Pennyroyal Plateau) for this purpose, but the area had not yet been surveyed. So, Virginia asked for more time to consider the matter:[20]

> That in case the quantity of good Lands on the South-East side of the Ohio upon the waters of the Cumberland River and between the Green River and Tennessee River which have been reserved by Law for the Virginia troops upon Continental establishment should from the North Carolina line bearing in further upon the Cumberland Lands than was expected prove insufficient for their legal Bounties the deficiency should be made up to the said Troops in good lands to be laid off between the Rivers Scioto & little Miami on the North-West Side of the River Ohio in such proportions as have been engaged to them by the Laws of Virginia.

Here was the vindication of Christopher Gist's first journey to the Ohio country, a public proclamation of the value of the lands that Gist had so accurately described. Virginia's western land claims included all of Kentucky and everything north of the Ohio to the drainage limit of the Great Lakes. Virginia might have chosen to retain the Wabash Valley, or the Illinois, but those lay too far distant. It might have chosen the woodlands between the Great Miami and the Little Miami, as John Cleves Symmes of New Jersey was about to do, or the southeastern quarter of Ohio, which lay far closer to settled territory and which Manasseh Cutler's Ohio Company of Massachusetts would take.

What Virginia chose, instead, was the best tract of potential agricultural land anybody had seen, the rolling prairies and oak groves that lay between the Little Miami and the Scioto which, although it did not include the Pickaway Plains on the east bank of the Scioto, nevertheless encompassed within an irregular polygon nearly everything Christopher Gist had written about so favorably thirty years before. The land had not been forgotten in Virginia, only its occupancy postponed. Now it would be officially designated as the Virginia Military District, open to settlement by anyone who had received, either through entitlement from military service or subsequent purchase, the land warrants issued to troops of the Virginia line.

The interest Virginians had in attractive land was shared by their North Carolina neighbors. Richard Henderson bypassed the governments of Virginia and North Carolina and purchased much of Kentucky and Middle Tennessee from the Cherokees "for a cabin full of goods" in 1775, at a time when settlements west of the Appalachians were forbidden. He called his great scheme "Transylvania." Virginia denied its legality and set aside the Pennyroyal/Barrens portion of Kentucky as the original Virginia Military Reserve.[21] Undaunted, Henderson turned his attention to the Nashville Basin, which he presumed would not fall under Virginia jurisdiction, and in 1778–1779 his company founded the settlement that would become Nashville.

Both Virginia and North Carolina had designs on the Pennyroyal, but its status was uncertain because the Virginia-North Carolina border (now, Kentucky-Tennessee border) had not yet been run that far west. In 1779 Thomas Walker was appointed to survey the boundary extension for Virginia, and Henderson himself took responsibility to run the identical line for North Carolina. The two parties soon ran into conflict, with Henderson's group claiming that the Virginians were veering too far to the south. When they reached the Pennyroyal/Barrens (which the laconic Walker noted only as having "little or no timber in it, in many places"), Henderson was relieved to discover that the Nashville Basin would not fall in Virginia.[22] Thus Virginia and North Carolina would share the Pennyroyal; both states recognized the quality of its land and both established military reserves there, thus creating a market for land.

It seems that Virginia thought the entire Pennyroyal/Barrens would lie north of the line, perhaps even that the Nashville Basin would as well, which explains the reference concerning the "North Carolina line bearing in further upon the Cumberland Lands than was expected." In any case, it provided a pretext to retain the Little Miami-Scioto tract, although not one that was above suspicion in the minds of those envious of Virginia's already large holdings. In March 1788 the congressional committee revising the land ordinance recommended that the size of the Virginia Military District in Ohio be limited to the lands within twenty-five miles of the mouth of the Scioto.[23] That restriction would have defeated the purpose of choosing the tract, since the southern, unglaciated, hilly end along the Ohio River was the only poor land in the whole reserve, as Gist had well described in his journal. In July 1788 a resolution was passed, with only the western land states of Virginia and Connecticut voting against it, that withheld any Ohio land from Virginia until the "deficiency" in Kentucky could be determined.[24] Finally, on August 10, 1790, the Congress of the United States passed an act officially opening the tract between the Little Miami and the Scioto to settlement, several years after settlement had actually begun.

Although the details of Virginia's claim north of the Ohio River were, for all practical purposes, left at the limits suggested in 1780, the larger issue of western land claims of the various states lingered until March 1, 1784, when the conditions of Virginia's cession of all its western lands were resolved in principle. Immediately following that question the committee chaired by Thomas Jefferson presented its report creating the Northwest Ordinance.[25] The way was being cleared for a new system for dealing with what came to be known as the public lands. Schemes such as Virginia's Ohio Company and Loyal Company, along with Henderson's Transylvania and numerous other putative colonies like "Vandalia" and "Indiana" (both projected to cover parts of what are now West Virginia and Kentucky) ultimately failed. But it was still possible for an individual to petition Congress and receive a substantial grant of land.

On August 29, 1787, John Cleves Symmes submitted a petition in which he asked for a grant "on terms similar to the grants of July 23 and 27, 1787 of Sargent and Cutler" (grants made to the Ohio Company of Massachusetts along the Muskingum) for the area between the Great Miami and Little Miami rivers in southwestern Ohio.[26] Symmes had been elected a delegate to Congress from New Jersey in 1785, and thus he was familiar with the procedure for obtaining lands. Later known as "Judge Symmes," for his appointment as one of three initial justices of the Northwest Territories, he had been a respected figure in New Jersey. Born on Long Island of New England parentage, he had moved to Sussex County, New Jersey, in 1770 and during the Revolutionary War had been placed in charge of various military forts on New Jersey's frontier.[27]

His association with the Miami Valley apparently was due to the enthusiasm of Benjamin Stites, an Indian trader from New Jersey, who impressed Symmes with accounts of the land found between the Great and Little Miami rivers. Symmes visited the Ohio Valley in 1787, but upon returning to New Jersey he first attempted to promote a land scheme in the Wabash Valley, north of Vincennes. It is not clear whether Symmes actually saw the Wabash Valley, but he did visit southwestern Ohio. His Miami petition received favorable action in Congress, and in November 1787 Symmes began to promote his scheme:[28]

> It lies in north latitude thirty-eight degrees, and the same with Virginia. . . . The land is generally free from stone and a rich, easy soil for tillage. There are no mountains and few hills, so that the country for the most part is level! . . . The finest timber of every kind known in the middle states, with many other sorts of more southerly production, grow in plenty here, but there is very little underwood or brush.

As much could have been said for other sections of the middle Ohio Valley. The prairies between the Great and Little Miami rivers that Gist had seen lay north of the Symmes tract, and hence they were not mentioned in the circular. Perhaps Symmes had not seen the prairies. Or, if he had gone that far inland, perhaps he was so unaccustomed to grassy lands that he was unable to recognize their value. The uplands along the lower Miami rivers that Symmes described were fairly level, although the interfluves were dissected by some deep valleys cut down into the limestone layers beneath the glacial mantle. There was scarcely a spot in the Miami Valley where anyone could have stood and not seen a ridge crest nearby terminating one of the long upland slopes that surround the lower, more steeply sloping valley segments. There were extensive bottomlands along the Great Miami, although they covered only a portion of the area. Partially forested, rolling, dotted with numerous small clearings and Indian old fields, it lacked the Virginia District's feel of expansiveness. It was the sort of country that had "yeoman farmer" written all over it.

Symmes's Miami Valley was not, to appearances, in a class with the Nashville Basin, Pennyroyal, Bluegrass, or Virginia Military District, but it would develop rapidly, and its agriculture would greatly stimulate development of the Corn Belt. The land that Symmes selected was better than that chosen previously by the Massachusetts-based Ohio Company along the Muskingum, but inferior to that claimed by Virginia. The parcel Symmes took was "the most eligible one that remained," at least in terms of fronting on the Ohio River and lying upstream from the falls at Louisville.[29] Both the valleys of the Muskingum and the Great Miami would have looked familiar to a New Englander or to one, like Symmes, accustomed to land no better than what Long Island or the hills of northern New Jersey had to offer. But woodland slopes held less interest for men accustomed to the limestone lowlands of Pennsylvania or the Great Valley in Virginia.

The lands that Virginians chose are instructive, especially because Virginia constantly pushed the limits of its territory and in nearly every case exercised the advantage of priority in making the first choice from among the blocks of land that were available. Whether Virginians had a special sense of the soil that allowed them to discern the good lands from the indifferent is open to question. Stewardship of nature was not a remarkable trait among those men whom Beverley described as "masters of great tracts of land." The problem of worn-out and abandoned lands already existed in Virginia and the Chesapeake by the time the country west of the Appalachians was being scouted by Thomas Walker and Christopher Gist. More likely, the choices were made in terms that would have been shared with the other seaboard colonies or any other European-derived population. If the choice in lands was undertaken with a view of promoting future developments, secured by a vision of commercial agriculture, the hilly backwoods country was not an inviting prospect.

Gist's remark written on the Ohio prairie in 1751—that it "wants nothing but Cultivation to make it a most delightfull Country"—probably gives the best insight of all. The lands perceived to be the best were those that had already been modified by human activity. They were, in another terminology, disturbed lands: stretches of treeless prairie, cleared bottomland, or open woods that had supported people for centuries before Euro-Americans came to know them. Woodlands were given favorable comment if they lacked an undergrowth, a condition that might be natural—a "climatic climax," to use an old term from plant ecology—but more likely a sign of human modification. The prairies and champaigns that attracted the French were identical to the meadows and openings seen by the English. And the farther to the east and to the south these exceptions to forested conditions were described, the more certainly they were a product of how the land had been used.

The label "disturbed land" carries with it a negative connotation that should not be there. "Improved land" would be a better term, because that is the point of making the distinction. Land that has been deliberately altered by some form of human activity carries no premium of recognition in a society that views any such impacts negatively. But today's environmental ethic, sensitive as it should be to deleterious human impacts on nature, offers the wrong perspective if the clock is turned back to the late eighteenth-century trans-Appalachian West. Aboriginal North Americans, like people everywhere, had deliberately disturbed their surroundings in order to produce an increase in nature. Burning to increase the grasses and the numbers of animals who fed on them was the most important modification in terms of area affected.

There were other ways of modifying the environment that led to an even greater increase in its sustenance for humans. Localized clearing, weeding, and planting were to the riverine habitats what massive burning was to the uplands. It was in the bottoms that the Indians raised their crops, and there, too, would the white settlers who replaced them. The early history of the Corn Belt of the Middle West would become a record not only of pioneering new lands but also of retaking used lands. The Indian tribes driven out of their former homes by the armies of white soldiers were not even the first occupants of the land in most localities. The record of who first undertook the work of improving the environment to sustain human populations stretches far back into prehistory where it becomes text for archaeological investigations.

The crop most often planted in the cleared bottomlands was corn, the staple of Native Americans and the common grain brought west of the Appalachians by early migrants coming out of the great swath of farming country from Pennsylvania to the Carolina Piedmont. Statistics on its production in the U.S. Census through 1890 are found under the label, "Indian corn," distinguishing it from the British "corn" (wheat or oats) and simultaneously acknowledging its New World source. But long before 1890 the corn that was being grown on thousands of farms in mid-America was no longer the corn which the Indians at Jamestown and Plymouth had shown European colonists. The spread of farming from the eastern valleys to the five islands west of the Appalachians involved not only the migration of people, but also the diffusion and recombination of races of maize. The evolution of Corn Belt Dent corn joins the record of prehistoric North American agriculture with that of the farming practices of Upland Southerners. It was the Southerner's reliance on corn as a feed grain for fattening livestock that projected the crop's utility ahead of any other and thereby provided the foundation of Corn Belt agriculture west of the Appalachians.

4

Zea Mays

THE FIRST EXTENSIVE exploration of the North American mainland was made by Hernando de Soto's expedition, which landed at Tampa Bay in May 1539. De Soto had equipped his party of some 650 Spaniards well, including a store of five thousand bushels of maize and a herd of perhaps four hundred hogs. Both maize and hogs were loaded aboard in Cuba, the hogs having been introduced there earlier from Spain, the maize from Mexico. Had de Soto wished to commence swine/maize agriculture immediately he could have done so, although that was not, of course, the object of his expedition. Nor was maize then a feed grain for hogs. New World Spaniards used it as their principal bread grain, after the fashion in Mexico. The hogs were a long-snouted import from Extremadura, on the Portuguese borderlands of Spain, where de Soto and many of his men had lived. The animals were accustomed to feeding on mast in the woodlands and to being driven by swineherds.[1] De Soto's hogs traveled well, and they multiplied rapidly. Eventually, some of them found their way into the Indian villages, where they were tended and soon became a favorite food item. Feral animals became the razorbacks of the southern uplands.

The great supply of maize seems to have been abandoned soon after the trek north across Florida began, however. The easier practice of taking food from the Indians was adopted (often in their absence, because once advance word of de Soto's party reached a village, the inhabitants fled). Maize was abundant nearly everywhere the expedition went, as were plentiful crops or stored quantities of beans and squash and many varieties of fruit. De Soto and his party ate well, but it was riches they sought. His expedition marched in search of the great treasure they never found, from the Florida peninsula north to the vicinity of Chattanooga, south through the Coosa Valley of Alabama, then north again to cross the Mississippi and follow the Arkansas River to Little Rock before heading south once more.[2] They crossed, or at least touched, all of the southeastern states, from Texas to South Carolina. Nearly everywhere they went there was an abundance, to which the party usually helped themselves. The greatest abundance was reported in maize, whether in the store of it, the size of the crop in the field, or its consumption.

Maize was a valley or lowland crop. Those who planted it commonly had villages either near the fields on bottomlands or on nearby terraces or bluffs.

Near the southern bend of the Tennessee River in north Alabama the fields were continuous between villages; the valley was a permanent, settled home of agriculturalists when de Soto was there. Indians grew several crops of maize annually, staggering the dates of planting, so that it would be available over a long season. As Marquette described an Arkansas village in 1673, "we saw at the same time some that was ripe, some other that had only sprouted, and some again in the Milk."[3] The sizes of the villages were roughly correlated with the amount of agricultural produce in the nearby fields, with the greatest intensity of agriculture and complexity of settlement along major rivers. The Coastal Plain's deep Mississippi embayment, stretching north to the southern tip of Illinois, had large crops of maize; the lower Ohio, Illinois, Wabash, and Missouri valleys were occupied in similar fashion at that time.[4]

Maize had been grown for hundreds of years in the valleys where de Soto observed it, but it was by no means the first crop grown by the inhabitants of eastern North America. The archaeological record contains numerous examples of plants that, once entered, whether by natural spread or human transport, underwent alterations in their usable parts, which would indicate deliberate human selection for improvement. One of the first plants that can be identified as a subject of human care in eastern North America is a small gourd, *Cucurbita pepo*, that probably originated in northeastern Mexico. It appeared in the lower Illinois valley seven thousand years ago. *C. pepo* was not a domesticate at that early date, but rather it survived naturally in clearings once introduced. The bottle gourd, *Lagenaria ciceraria*, arrived from Mexico and was grown by the inhabitants of the Pomme de Terre River Valley of western Missouri four thousand years ago.[5] Selection for thickness of rind, or shape and size of seeds, eventually produced the variety of squashes and gourds grown in eastern North America.

The deciduous hardwood forest established across the mid-continent following the Pleistocene was an early source of food. Charred hulls of pecans, *Carya illinoiensis*; hickory nuts, *C. ovata*; black walnuts, *Juglans nigra*; butternuts, *J. cinerea*; and acorns of oak trees, *Quercus* spp, are dated in Mississippi Valley archaeological sites between 9,000 and 8,000 B.P.[6] Burning removed undergrowth to make nut collection easier, and deliberate weeding of unwanted competitors of the desirable species helped increase the harvests. The transition from nuts to seeds as a major dietary component began in the Mammoth Cave region of Kentucky about four thousand years ago.[7] The shift was also marked by a transition in the landscape. Nut-gathering took place in a woodland setting, but the environment became substantially more open once inhabitants learned the ways of planting.

The plants dating before A.D. 1, both native and tropical, grew at a variety of sites associated with some type of human habitation. They were to some extent "camp followers" that could have first appeared on disturbed ground following occupation. Undemanding as to site, they survived untended as well as in a cultivated condition. Common Sunflower, *Helianthus annuus*, may have been domesticated as early as 4,000 B.P. Goosefoot, *Chenopodium berlandieri*, a weed that produces starchy seeds, dates from about the same period. Knotweed, *Polygonum erectum*; Maygrass, *Pharolis caroliniana*; and Little Barley, *Hordeum pusillum*, are three other plants producing starchy seeds that have been identified in the archaeological record of sites in the mid-continent region about two thousand years ago.[8] These were native weeds that probably spread from the lower Mississippi Valley, but their earliest dates of apparent cultivation are later than those for the more exotic *C. pepo*.

Wild grapes, *Vitis* spp, probably received more favorable comment from the early Spanish, French, and English explorers than did anything else they saw in the American landscape. Grapes grew in enormous bunches from vines that climbed the wooded margins of clearings, cultivated fields, or other openings. The French, especially, were enthusiastic about the possibilities of viticulture. Other fruits, such as plums, *Prunus* spp, were consumed by many a European feasting at an Indian's table and typically received favorable comment, eaten either fresh or dried.[9] These were the fruits of nature, to be sure, but of a nature that had been tended to a state of advanced productivity. Fruits of the woodland and the woodland margins, enhanced by humankind like the prairies, impressed Europeans that this was a bountiful land.

Maize and beans were comparatively late arrivals in eastern North America. Maize, *Zea mays* L., was not certainly present east of the Mississippi River until A.D 600; beans, *Phaseolus vulgaris*, arrived after A.D. 800. With those two additions, the "trinity"—corn, squash, and beans—was complete in the East.[10] They were raised together, often in the same fields, and taken as a group provided complementary amino acids for a complete vegetable protein diet. The trinity was adopted by Europeans from New England to the southern colonies as soon as contact with the Indians took place. In addition to their well-known reliance on corn, Southerners also retained use of some of the early weeds. Purslane, *Portulaca oleracea*, a camp follower that came from Mexico along with maize around A.D. 800, carpeted the buffalo traces; it was (and still is) harvested for greens. Knowledge of the herbs used for potions and teas and the ways of preserving the fruits of vine and tree also appear to have been transferred upon contact.

Two hundred different varieties of *Zea mays* were being grown somewhere in the Americas, from eastern Canada to Argentina, at the time of Columbus's first voyage.[11] The great genetic variety that characterizes maize is at once a clue to understanding the plant's success in adapting to new environments and a shroud around the mysteries of its origin. The most parsimonious of the theories of maize origin holds that the plant is a human invention.[12] It was likely derived in several steps from a wild plant, teosinte, *Z. mexicana*, a native of the highlands of southern Mexico, beginning at least eight thousand years ago. Human selection of a mutation of teosinte that produced a more easily accessible fruit may have been the first step. That was followed by subsequent selection for ears with exposed kernels, making shelling easier. Selections for size of ear and quantity of grain led to numerous strains which crossed freely with one another as well as with the wild teosinte.

Maize first emerged in highland Mexico or Guatemala, a tropical latitude with relatively constant day-lengths. This circumstance preadapted maize for spread into tropical latitudes of northern South America, although diffusion into the long day-lengths of middle-latitude summers required an additional period of adaptation. The date when maize began its northward spread from the highlands of Mesoamerica is unknown, but the progress was gradual, through contact of one band of people with another, rather than in the manner of rapid, direct transfer of the sort that would accompany long-distance migration. Maize entered the southwestern United States around 1,000 B.C., where it was incorporated gradually into the food complex of hunter-gatherer societies.

One of the early varieties of maize was a popcorn known as Chapalote, which has twelve rows of small, globular kernels (fig. 10). Crossed with a primitive eight-row type known as Harinoso de Ocho, it produced a hybrid, Pima Papago. The variety that diffused eastward from New Mexico was Pima Papago's eight-row derivative, Maiz de Ocho, which incorporated the more primitive eight rows with an apparently evolved trait of large kernels. Maiz de Ocho has been documented in the Rio Grande Valley near Albuquerque about A.D. 370.[13] From the upper tributaries of the Rio Grande it could have spread equally well into the valleys of the Canadian, Red, Cimarron, or Arkansas rivers, although the exact route of its spread across the Great Plains is unknown. Nor have other paths of diffusion been ruled out, such as the more direct route from Mexico across the Gulf Coastal Plain of Texas, the "Gilmore Corridor" espoused by some archaeologists.[14]

It is certain, however, that maize had to have been planted, harvested, and replanted at regular intervals in order to be maintained. Unlike wild plants that have some means of dispersing their seeds to ensure future reproduction, kernels of mature corn remain tightly bound to the ear and rarely fall in a manner that would allow individual seeds to take root the following year. If the crop was abandoned, nature might allow a feral derivative to survive for enough generations to permit later redomestication, although that would presume that the weather cooperated and that animals did not consume the fruit. Circumstances suggest that the dispersal of maize was largely a function of human migration or trade.

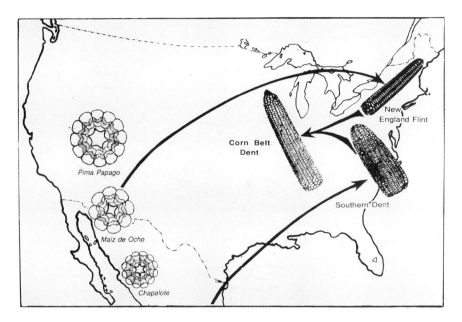

Fig. 10. Races of maize and their diffusion. From Walton C. Galinat, "Domestication and Diffusion of Maize," Richard I. Ford, ed., *Prehistoric Food Production in North America* (Ann Arbor: University of Michigan Museum of Anthropology, 1985), fig. 8.7. Reproduced with permission.

The likely route of eastward movement across the Great Plains was the alluvial bottoms along one or more rivers, any of which would have led in time to the Mississippi. Maize would not have been confined to lands along the rivers once the humid lands of Texas, Oklahoma, or Missouri were reached, although floodplains still remained the favored sites. Despite its high nutritional value and ease of incorporation into existing agricultural systems of the mid-continent, increased maize production was most often associated with demographic changes and fluctuations in the rest of the subsistence base. Its mere availability did not mean that people turned to maize to the exclusion of other dietary inputs.

Although the ancestral Harinoso de Ocho carried genes preadapting Maiz de Ocho to highland growing conditions, maize spread to the east (and west) more rapidly than it did to the north. Maiz de Ocho had to keep reincorporating its Harinoso de Ocho ancestry through introgressive hybridization, gradually moving toward northern latitudes of cool spring conditions and long summer days. Maize growing appeared among the Hurons, in the area between Georgian Bay and Lake Simcoe, Ontario, around A.D. 1000. The crops were by no means marginal in this zone of only 140–160 frost-free days per year. Samuel de Champlain's 1616 estimate of the Hurons' total population at 30,000 and the assumption that maize was their staple food together suggest fields perhaps on the order of thousands of acres. The Hurons acquired maize first, then sunflowers (A.D. 1300), and finally beans and squash (A.D. 1400).[15]

The common name for Maiz de Ocho in the United States is Flint corn, of which there were several races, including the Northern or New England Flint, a name suggesting the northern latitudes in which the plant was commonly grown. New England Flints were grown from Atlantic Canada to North Dakota, but also nearly everywhere else that corn grew in the northeastern United States at the time of European contact. Tropical Flint corns dominated in other areas. The typical Northern Flint variety had the smooth, rounded kernels of its Chapalote ancestor and had eight rows of kernels on a long, slender, tapered ear (fig. 11). It, or its ancestors, spread from highland Mesoamerica to northern Mexico, up the Rio Grande Valley of New Mexico, and followed unknown routes across the Great Plains, becoming established as a major food crop east of the Mississippi River more than half a millennium before it was recorded by Europeans.

There was another type whose history is poorly known but which contributed the additional germ plasm necessary to produce the crop that came to characterize the Corn Belt. Probably in Mexico there developed a race of corn with dented, rather than smooth, kernels. There is scant archaeological evidence of its past, but it is known to have existed before A.D. 1500. The dented kernels are but one difference, however. Dent corn has a short and thick tapered ear, its plant has many tassel branches, and the ear has between fourteen and twenty-two rows of dented kernels (fig. 12). These physical differences, plus other genetic dissimilarities, caused William L. Brown and Edgar Anderson to exclaim that the Dents and Flints were "so different that, were they wild grasses, they would be considered as totally different species and might well be placed in different genera."[16]

The first reference to Dent corn in the United States was made by Robert Beverley of Virginia, who wrote in 1705 of the "four sorts of Indian corn" that grew there, two of which ripened early, two late. Of the late ripeners, Beverley wrote "one looks as smooth and as full as the early ripe corn, and this they call flint corn; the other has a larger grain, and looks shriveled, with a dent on the back of the grain, as if it had never come to perfection; and this they call she corn." He went on to claim that the Dent, or "she corn," is "esteemed by the planters as the best for increase, and is universally chosen by them for planting." John Banister of Virginia explained that the "she-corn . . . is more soft & feminine, on whose Superfices Nature has impress'd the Signature ♀."[17]

The close resemblance between the Southern Dents of the United States and a Mexican variety, *tuxpeno*, suggests that Mexico is the origin, although

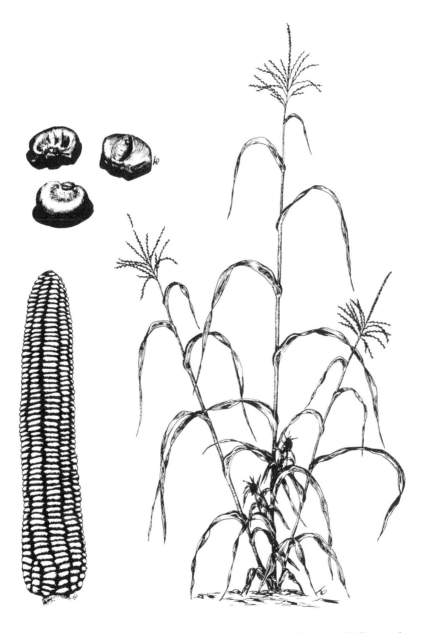

Fig. 11. New England Eight-Rowed Flint Corn. From Henry A. Wallace and William L. Brown, *Corn and Its Early Fathers* (East Lansing: Michigan State University Press, 1956), fig. 9. Reproduced with permission.

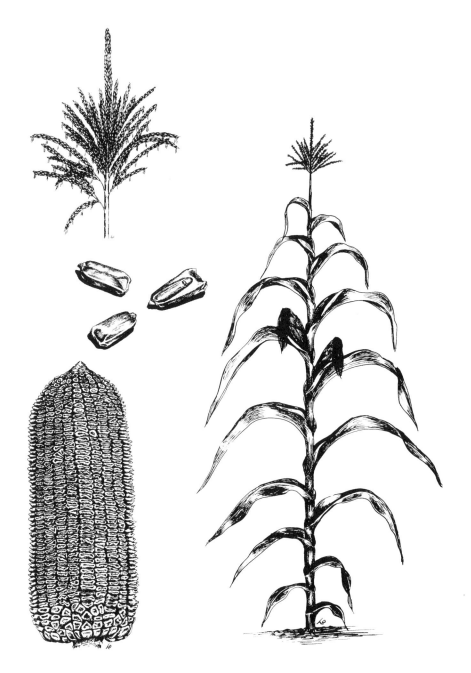

Fig. 12. Many-Rowed Gourdseed Corn. From Henry A. Wallace and William L. Brown, *Corn and Its Early Fathers* (East Lansing: Michigan State University Press, 1956), fig. 10. Reproduced with permission.

perhaps not the highlands of southern Mexico. Dent corn was growing in Mexico at the time of the conquest, as suggested in pre-Columbian art incorporating the dented kernel.[18] Wherever Dent corn was being grown, the Spanish found it and took it to the Antilles. Perhaps they, like the Virginia planters, recognized Dent corn for its productivity. Those five thousand bushels of maize that de Soto brought to Florida from Cuba may have included Dent corn, but that is not known. There is circumstantial evidence for the transfer of Dent corn to Virginia from Cuba, however, because Havana was the port of origin and outfitting for the Spanish expeditions to Florida and the southeastern Coastal Plain of the United States.

In the 1560s Spain established a Jesuit mission on St. Helena Island about fifty miles up the coast from the Savannah River. It was the most successful Spanish settlement at that time on the North American mainland. St. Helena had some agricultural activity, although the post was supplied largely by ships from Cuba. In September 1570 a party of Jesuits attempted to extend their reach northward by establishing a mission along the James or York estuary on Chesapeake Bay in order to convert the Indians in the vicinity of Powhatan's village. But the missionaries met a hostile reception and soon learned they were in danger. All of the Jesuits were murdered by the Indians the following February, and a relief ship just arriving with supplies was attacked and forced to turn back before it could land.[19]

The ill-fated Jesuit mission could have brought Dent corn straight to Tidewater Virginia. Agriculture on St. Helena may have incorporated the crop as well. Or, later Spanish trading missions may have brought it to the Gulf and Atlantic coasts. The references to Dent corn by Beverley and Banister are apparently the earliest clues of its presence in Virginia. Anderson and Brown's belief that the Dents included "some germ plasm exhibiting Caribbean influence" also suggests Spanish provenance but does not suggest where the contact took place.[20] The presence of Dent corn in early Louisiana similarly can be interpreted in terms of Spanish introduction.

Beverley anticipated the inevitable crosses between Flints and Dents when he added to his observations on planting the two together, "this also produces the flint corn, accidentally among the other."[21] Various types of Flint corn were widespread well before European contact, but the Dents seem to have spread from one or only a few points of entry. Because the French and Spanish accounts describing Indian corn production in the Mississippi Valley do not mention this distinctive ear with dented kernels, and given the keen interest travelers had in plants and their culture, it is possible that Dent corn simply wasn't present where they were. The Jesuits of Tidewater Virginia did not survive to describe their activities.

According to Anderson and Brown, mixed plantings of Flints and Dents were first described by John Lorain in a letter he wrote to the Philadelphia

Society for Promoting Agriculture in 1812: "by forming a judicious mixture with the gourdseed and the flinty corn, a variety may be introduced, yielding at least one third more per acre."[22] "Gourdseed," which has nothing to do with gourds, was one of the common varieties of Southern Dent corn. Lorain lived in central Pennsylvania, recently removed there from the vicinity of Philadelphia. He may be the same John Lorain who was listed as a planter on the Eastern Shore of Maryland in the census of 1790. Lorain freed his slaves and moved to Pennsylvania, where he was raising corn and fattening steers by 1809.

Gourdseed or Shoepeg (another variety of Southern Dent corn) could have spread north to the Eastern Shore and become established by the time Lorain lived there. Trade and migration routes would suggest a spread inland at least to the Shenandoah Valley, north to the limestone lowlands of south-eastern Pennsylvania, and south to the Carolina Piedmont between 1730 and 1790 (fig. 13). From those places it is known to have been taken west to the Ohio Valley by frontier agriculturalists, but it would likely have spread in all other directions as well—against the flow of the westward-moving frontier but within the matrix of trading connections up and down the Atlantic seaboard.

The annual report of the U.S. Commissioner of Patents in 1848 included statistics on corn varieties and production for all the states.[23] White Flint, White Gourd, Yellow Flint, and Yellow Gourd appeared widely at that time, indicating that some dispersion and mixing had taken place, at least partly the result of the Commissioner of Patent's practice of making seeds available to those interested. The label "Yellow Dent" appears in the Indiana entry only, a possible precursor of the type that would come to be known as Corn Belt Dent. The 1848 statistics suggest no geographical limits on attempts to grow the many varieties; rather, there seems to have been a common interest in experimentation to increase yields. All of the farmers who wrote were understandably interested in finding a type of corn that would produce more, year after year, than that which they were then planting.

A fuller report in the same series for 1850 suggests how widespread the enthusiasm for corn production had become. Farmers' letters typically gave details of strategies they had found successful or unsuccessful. Homer Winchell of Shelby County, Missouri, reported, "the New England yellow flint corn has been tried and proved a failure." In Knox County, Tennessee, "large white corn [probably White Gourdseed] is most esteemed." Long-eared White Flint performed best on the eastern end of Lake Ontario in Jefferson County, New York. Dr. John H. Weir of Madison County, Illinois, reported the great corn production on the American Bottom (floodplain on the east side of the Mississippi River opposite St. Louis), stating "the yellow and a mixture of the white flint and gourd-seed are the best and most productive."[24]

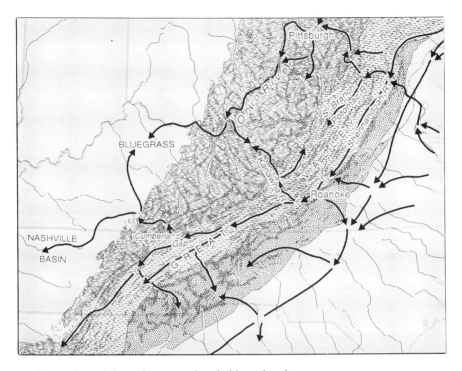

Fig. 13. Late-eighteenth-century Appalachian migration routes.

Corn Belt Dent—a cross of Northern Flint and Southern Dent—no doubt had numerous independent origins, perhaps as early as Beverley suggested in Virginia. It is plain that by the late 1840s, at least, the genetic foundation necessary to achieve the crossing of races of maize was widespread in the area that would become the Corn Belt. The significance of Corn Belt Dent is that it represents a geographical as well as a genetic joining of unlike ancestries. Corn Belt Dents were variable, open-pollinated varieties that had the potential to incorporate the heterosis (hybrid vigor) of recombining two races that had long histories of evolution separated from one another. The Corn Belt Dent type "tended to have one well developed ear, frequently accompanied by a small ear at the node below this primary one."[25] The well-developed ear contained a large, cylindrical red cob and between fourteen and twenty-two straight rows of yellow, dented kernels. The Southern Dent contributed the high row number, indentation of kernels, red cob, and general shape of the plant; Northern Flint influence was apparent in the shape of the cob and the configuration of the tassels.

Henry A. Wallace promoted a single-origin theory for Corn Belt Dent, and he gave most of the credit for early developments to one man. Robert Reid, who moved from southern Ohio to the lower Illinois River Valley in

1846, brought with him a late-bearing reddish-cob Dent variety he called "Gordon Hopkins" (after the man from whom he obtained the seed) that came originally from Rockingham County, Virginia, where it had been grown since 1765. Reid's crop of Gordon Hopkins did not germinate well in the Illinois spring of 1847, so he filled in, between the hills that had sprouted, the seeds of "Little Yellow," undoubtedly a Northern Flint. Thus, an accidental cross took place, one that produced better than Reid had thought possible. Further selection produced a cylindrical, ten-inch ear with eighteen to twenty-four straight rows of fairly smooth kernels. In 1893 Reid's Gordon Hopkins/Yellow Flint progeny won him a prize at the Chicago World's Fair.[26]

How much of the Reid story is true is difficult to say, although there is no reason to doubt the general sequence of events. The same process could have been repeated dozens of times in the first half of the nineteenth century, the "mixture of the white flint and gourdseed" in the American Bottom being yet another example. But placing Southern Dent corn in Rockingham County, Virginia, in 1765 has clear implications for its subsequent spread. Rockingham is in the Shenandoah portion of the Great Valley, one of the most important migration corridors of colonial and early postcolonial America. Those who passed through there would have included Pennsylvanians who later moved south to the Carolina Piedmont, Virginia planters moving up from the Piedmont or Tidewater, and many of the early settlers of Kentucky, Tennessee, and southwestern Ohio.[27]

From Rockingham County the seed could have been taken equally well to the Virginia Military District or the Miami Valley of Ohio, the Bluegrass or Pennyroyal of Kentucky, or the Nashville Basin of Tennessee—and beyond those areas, later, to the limestone lowland of southern Indiana, the Sangamon country of Illinois, or the Boonslick of Missouri. The earliest settlement of all of those areas was accomplished before Reid arrived in Illinois. Rockingham County probably saw as many of the important currents and cross-currents of westward migration as any other place that could be named prior to 1820. If Southern Dents were not already being grown in southeastern Pennsylvania by 1765, they were established in the crop cycles of Tidewater planters. Either way—through introduction from Pennsylvania or from the Tidewater—a date as early as 1765 for Dent corn in Rockingham County suggests that Dent varieties, whether Shoepeg, Gourdseed, or others, could have been taken into all of the early corn-producing districts that were to emerge west of the Appalachians.

The corn harvest of the United States in 1839 amounted to 378 million bushels, or twenty-two bushels for every person living in the United States at that time. Nearly one-fourth of the crop was grown in Ohio, Indiana, and Illinois, and as much more was produced in the two leading corn states, Tennessee and Kentucky. None of the eastern states, save Virginia and North Carolina, produced as much as any of the five western ones. No statistics on types of corn are available, although Tennessee, Kentucky, Ohio, Indiana, and Illinois all would have been introduced to the Southern Dents by that date, as Virginia and North Carolina had been; this much can be inferred from patterns of migration and the well-known tendency to transplant familiar agricultural practices to a new environment. And since the Flints were wide-spread, the possibility of accidental or purposeful crosses would have been present in any place reached by the tide of migration passing through the Shenandoah Valley.

Corn was raised in greatest abundance in 1839 in those states that had received the settlers most likely to have brought Southern Dent corn along on the trip west. If the patterns of migration to the frontier can also be presumed to be the paths of diffusion of the various cultivars of Dent corn, then the Great Valley heritage of the early Corn Belt becomes more sig-nificant. Regardless of whether the greater production was due to Flint-Dent crosses at this early date or whether greater yields came simply from planting the Southern Dents rather than the Northern Flints, the connection with Virginia (and, by extension, with Maryland and Pennsylvania as well) is en-hanced. If maize genetics was a major factor in establishing the Corn Belt as a region where farmers possessing Dent corn just happen to have settled, then explanations of the region's emergence based on climate and crop af-finities assume less significance.

The climate of the region from Tennessee to Ohio to Missouri did not retard the spread of Dent corn, but neither was it especially important, except in a benign effect that did not discourage an already common cultural prac-tice. The longer growing season required to bring Dent corn to maturity did place a northern limit on its culture, although that limit was not reached in the Middle West by 1839, nor even by 1859. Explanations of the Corn Belt that postulate an "orientation" to corn that certain regional cultures seem to have had also become less significant. Those who carried ears of Dent corn west across the Appalachians from the Great Valley probably were unaware that they had been favored by circumstances of past history—human as well as genetic—while others had not.

Of equal importance was the use that American farmers made of corn. It had been a food crop for the Indians, and it continued in that role for many Euro-Americans, especially in the South. But the Corn Belt emerged as a region where corn was raised primarily to feed animals, not humans. It was in its unique quality as a feed grain that the Southern Dent stood truly superior to any of the other varieties available in the early nineteenth century. Instead of the hard, flinty kernels used for meal or flour, Dent corn had softer kernels, more easily chewed by animals, and required no further pro-cessing. In fact, Dent corn did not even require harvesting, since animals could be turned loose into a field of mature corn in the fall to "hog down" the crop.[28] Dent corn outyielded the other varieties, its longer season to reach

Fig. 14. Cornfields on the Nolichucky River bottoms, Washington County, Tennessee. Early white settlers found large stretches of easily cleared bottoms and planted them in crops. Later, livestock were driven over the Appalachians through these East Tennessee valleys, and drovers purchased local corn to feed their market-bound stock.

maturity posed no problems, given the migration patterns of farmers emanating from the upper South before the Civil War, and it demanded far less labor to convert the product directly into animal fat and protein.

Planting a crop of corn was one of the first priorities of new settlers in the Ohio Valley.[29] No matter where the Euro-Americans who had learned the American Indians' maize culture went, they selected fields very much like those the Indians had chosen. Bottomland clearings were the favorite of the whites, the clearings themselves having been the work of Indians over the past centuries (fig. 14). The old fields commonly mentioned in the early journals were sometimes small clearings in a solid woodland, but many were extensive enough that the French and, later, the Americans, called them prairies, just like the far larger grassy expanses on the uplands.[30]

The bottomland prairies were more likely the result of the hard labor of girdling and deadening in the Indians' fashion, whereby trees were stripped of their foliage, branches, and even large limbs, the brush was set around the base of the tree, and all was ignited. The Indians grew crops of corn, beans, and squash in the deadenings, while the standing tree trunks slowly decayed. Eventually all signs of the once-heavy forest growth were eliminated. It was in such fields that prehistoric people slowly began to experiment with cultivation. Clearings allowed the light-loving varieties to increase, weeding eliminated their competitors. Maize was incorporated into the system in the first millennium A.D., and then the clearings grew larger, the cultivation more regular.

A frequent generalization made about the expansion of agriculture into the Middle West in the early to mid-nineteenth century concerns the relative avoidance of open tracts of prairie.[31] Turning the thick prairie sod was difficult until the self-scouring plow was available, so attention was diverted away from the fertile but practically unusable open stretches of sodgrass. The point is valid but ignores the alternative. A tract of forested bottomland would have been a far more daunting prospect to a corn farmer wishing to bring land into cultivation than would a tract of open prairie—it would, that is, unless somebody had already removed the trees.

Despite the many maps of vegetation that show the riverine bottoms of the Middle West as forested in their natural state, it was not the forest cover itself that attracted early cultivation. Rather, it was the smooth, fertile bottomland with the large trees already removed that offered the quickest path to agricultural success. Bottoms and terraces were preferable because they were no longer in a wooded condition, and they were found not just in patchy clearings, but in stretches fringing alluvial rivers both large and small.

A little more than one-third of the 1839 corn crop in Ohio, Kentucky, Indiana, Illinois, and Missouri was grown in the counties along the five principal rivers: the Ohio, the Wabash, the Illinois, the Missouri, and the Mississippi. Even in Kentucky, where cornfields spread across the inland Bluegrass section, about one-fifth of the 1839 crop was raised on bottomlands along the Ohio River. More than 60% of Illinois and Missouri corn was grown in the two states' river counties. While by no means all of the corn was planted along the major rivers, there were smaller streams in Illinois—the Kaskaskia, the Vermillion, and the Sangamon to name but three—that would have increased the proportion of the crop grown in bottomlands. The Salt River in Missouri, both Miamis and the Scioto in Ohio, the White in Indiana, and the Green in Kentucky are further additions to the list of smaller streams along whose courses large cornfields appeared in the early nineteenth century. In those cases, too, the most eagerly sought lands were those already cleared of heavy timber.

An early settler along the Muskingum River in Ohio described the history of one alluvial terrace:[32]

> The soil was a sandy loam. . . . This spot had probably, at a remote period, been cultivated by the Indians, as many such places are found at various points on the Muskingum, covered with a growth of saplings, while the adjacent lands are coated with forest trees. The autumnal fires of the Indians, followed up regularly for a long series of years, had prevented their

growth into trees, which had doubtless at some remote time covered it as well as the neighboring hills.

Thick bushes and some decayed tree trunks were all that needed to be cleared. No estimates of the amount of unforested—or at least easily cleared— bottomland have been made that would afford comparison with anecdotal evidence such as this. What can be inferred from the map of the expanding Corn Belt is that its rapid westward spread was produced by leap-frogging over nonriverine areas to occupy the best bottomlands, wherever they were found. The path of least resistance followed the rivers, avoiding both the uncleared forest and the upland prairies. Riverside farming offered both agricultural land and transportation access.

Corn was thus a bottomland crop for the early white settlers just as it had been for the Indians. In some cases the very fields used by the Indians were taken over, although the extent to which such a direct succession was possible depended on timing. In most areas there was a hiatus between the time Indian titles were extinguished and farmers arrived from the east. If the gap in time was long enough, the land reverted to forest. But in some cases the interval was brief, and the Indian old fields had no more than a light cover of shrubs and weeds that needed removal.

In 1819, two of the Renick brothers, cattle feeders from the Scioto Valley of Ohio, made a trip west seeking possibilities for relocating their operations to the Missouri Valley. About midway in their journey they paused on a bluff overlooking the American Bottom of the Mississippi opposite St. Louis. The entry in their journal reads, "the greater part of the Bottom . . . is prairie of the finest quality and wants nothing but industry and art to afford some of the finest farms that any country can boast of." Twenty-six years later the following reference to the American Bottom appeared in an Alton, Illinois, newspaper: "From Mr. A. B. Davidson's residence on the bluff, on the road from Alton to Edwardsville, there may be seen, without moving from the same spot, by looking only in two directions, fields of standing corn the probable yield of which is estimated at 1,500,000 bushels."[33]

The Renicks would not have been surprised to learn that it took only a generation to see the development they anticipated blanket the whole area. The two men were already familiar with such a transition in their own Scioto Valley. The Pickaway Plains, along the east bank of the Scioto were "destitute of trees" when the first white settlers arrived in the late 1790s, less than forty years after Christopher Gist had seen it next to the Indian village. The black soil, eighteen to thirty-six inches deep, required the strongest teams of draft animals to break.[34] It was plowed, planted to corn, and has remained a gigantic cornfield ever since (fig. 9).

5

The Feedlot

THE CORN BELT is at once the most typical American region and the most productive. Its abundance—overabundance, even—somehow makes it more typical, a fulfillment of the true bounty of the American land. The Plantation South produced a large output, but its social system made it a national disgrace. New England, New York, and the Great Lakes region are normally found praiseworthy on cultural grounds, but the land has never been as productive. The Arid Plains, the Rocky Mountains, and the Far West are atypical on all grounds. But the Corn Belt of the Middle West is the Heartland.[1] Its very middleness has long caused pundits and scholars to seek within its bounds the typical in American life. The middle is the average, the average is typical of the whole, yet the average is also found to contain the very best the nation can offer. Where in America would one expect to find a typical family farm, with barns and cows and chickens and fields of golden grain? Where would one seek a typical community, with shops and schools and churches and clean, frame houses? Where would one expect to find that national values had achieved their highest expression, in the greatest good for the greatest number of people? The answer, in all cases, is obvious.

The Corn Belt myth is a powerful one, yet within it are some themes that help frame an understanding of the region's evolution. For example, there has long been confusion as to whether the Corn Belt is an extension of the North, the South, or both. Its middle location would suggest "both," although some authors have been at pains to describe the nature of the blend. In *Planting Corn Belt Culture*, written soon after Edgar Anderson had explained the historical significance of Flint-Dent hybrids, Richard Lyle Power asked, "has the same superior quality been attained by the blending of the two peoples?" An "interesting question for speculation," he concluded.[2] Power cast his analysis in the contrast between Yankees and Southerners. The Corn Belt was their supposed mingling ground where cross-fertilization could take place, yet Power, like others, was unable to identify where—or if—the blending occurred.

Avery Craven saw three elements contributing to the region (the third being foreign-born). From the Middle States came "sallow lean woodsmen . . . with homely vices and virtues and more than their share of half-starved emotions—men who without show or complaint set about reproducing the life they had known in such varied places as tidewater Virginia, back Pennsylvania, and the upper Carolinas." New Englanders and New Yorkers came later, "broke the prairie soils," and "added a bit more of professional service than was common to their neighbors."[3] These are durable stereotypes that have been embellished many times. A history of Knox County, Ohio, published in 1862, contained a pageant-like description of the county's settlement: "While [the] plain men from Virginia, New Jersey, and Pennsylvania are preparing their cabins . . . and making little clearings, a stray Yankee, solitary and alone, with a speculative eye and money-making disposition, is with pocket compass taking his bearings through the forest, soliloquizing about the chance of making a fortune by laying out a town." The editor of the *Chicago Weekly American* had a more succinct description in 1837. Watching the new arrivals from New York stream through the city, he wrote: "Yankees—a shrewd, selfish, enterprising, cow-milking set of men."[4]

Simplified versions of these stereotypes are embedded in the Corn Belt myth. Those from the South were daring and shifted for themselves, but were of low caliber. They were the woodsmen who swung the axe, felled the trees, and cleared the land for planting. Yankees lived by their wits, got others to work for them, and generally lent a hand organizing everything. The great fecundity of the soil, combined with plenty of rain and sunshine, seem to have provided all else. Since the Southerners came first, it was they who clung to the wooded prairie margins; the Yankees, arriving later, had no choice but to settle the prairies—which proved to be excellent for agriculture despite the lack of timber.

The land system created 160-acre farms, a leavening as well as a leveling. Expectations were as modest as the circumstances of the pioneer. No one, apparently, aspired to more than a good, comfortable life of husbandry. About the only disruption was the building of railroads that hauled the produce to market, enabling farmers to sell more and devote less time to subsistence. The region thus evolved, a landscape checkerboard of mixed grain-livestock family farms, peopled by a blend of folks from the North and the South who contributed their diverse ways to a new and stable synthesis.

As for the land and its condition at the time of settlement, the myth is easily dispelled. The early centers of the Corn Belt, from Tennessee to Ohio, were built around lands that for centuries had been altered through prior use by the Indians, centuries more before that by prehistoric people whose habits are known to some extent from the archaeological record. Whether prairies for grazing or bottomlands for corn, these lands demanded as little work to bring into bumper-crop production as any encountered in the course of westward settlement expansion in the United States. The Indians could have done no more had they left signboards, "Cropland" or "Pasture," for the whites to read. The woodsmen were real enough, but the backwoods frontier was not the Corn Belt.

The question of population sources is more complex. Upland Southern-ers—referring to people who hailed from the stretch of territory from south-eastern Pennsylvania to north Georgia, embracing the Piedmont on the east to the Interior Low Plateaus on the west—certainly did populate the Corn Belt. In fact, they provided by far the largest share of its early inhabitants. It was, rather, the backwoods ways of subsistence living associated with the label "Upland South" that were hard to find wherever Corn Belt agriculture emerged in the Middle West. The Southern influence was not limited to the traditional Upland segment, however, as would be revealed in sectional feel-ings that emerged both before and after the 1860s. If the term "Yankee" is applied only to those with New England/New York roots, which is its least confusing usage in a regional-cultural sense, then it is clear that Yankees played only a minor role in populating the Corn Belt of the nineteenth cen-tury and were perhaps even less influential in shaping the course of its agri-culture.

Most complex is the question of scale, the place of the yeoman farmer, and the "typical" family farm that is supposed to have stamped the region more than it did any other. James Lemon's *The Best Poor Man's Country* finds in early southeastern Pennsylvania a prototype for what might be termed middle-scale agriculture. Pennsylvania farmers raised corn, cattle, and hogs, but they relied mainly on wheat for the export market as cash income. They used the land extensively and were content to hold their inputs of capital and labor at a modest level. Lemon wrote, "at bottom, extensive farming was the result of the satisfaction of the average farmer and his response to economic conditions. He produced enough for his family and was able to sell a surplus in the market to buy what he deemed necessities."[5]

While there were doubtless thousands of Corn Belt farmers who held such a philosophy, that mode of thinking cannot compare with the scale at which corn-livestock farming was undertaken west of the Appalachians. To draw a parallel with cotton, sugar, or rice plantations of the South would be to err in the other direction, but there was a sense of increasing returns to scale built in to early Corn Belt agriculture for which there was no apparent prototype in early Pennsylvania. This has remained a feature of middle-west-ern agriculture down to the present day, dissonant at each stage of regional evolution with the supposed tradition of family farms and middle-scale, satisficing agriculturalists. Because the Great Valley and the Piedmont from Pennsylvania to Virginia provided the largest number of migrants to the newly emerging Corn Belt islands of the Ohio Valley at the beginning of the nineteenth century, and since Pennsylvania seems to offer little hint of where Corn Belt farmers acquired their fondness for doing business on a large scale, the obvious remaining choice is to seek a source in the influences coming from Virginia.

The Virginia Military District in Ohio was open to land entry by holders of certificates issued to Virginians who had given service in the Revolutionary War. This circumstance favored a large settlement from Virginia, although it did not cut off other sources of migration, because the land warrants themselves became a commodity that was bought, sold, and traded in the eastern states. Virginians of some means were especially attracted to the possibilities for stimulating the growth of this "New Virginia" between the Little Miami and Scioto rivers while they simultaneously made money in the land warrants market. The influence penetrating north to Ohio was not so much in the large numbers of people removing there from Virginia as in the overall control of the land business that was exercised by some key Virginians who inserted their own ideas about land tenure, society, and politics into the early phases of Virginia Military District settlement.

Nathaniel Massie, born in Berkeley County in the Shenandoah Valley of (West) Virginia, was the son of a prosperous land owner. Young Massie moved to Kentucky, and from there he began exploring the lands north of the Ohio River set aside as the Virginia Military District. In 1790 he was appointed deputy surveyor of those lands. In the next decade he would survey more than three-quarters of a million acres west of the Scioto River. Massie founded the town of Massie's Station (now Manchester) on the Ohio bottoms in 1791. In 1796 he laid out what would become the principal town of the Virginia District, Chillicothe, on the bottoms of Paint Creek near its junction with the Scioto, thus taking the Shawnee's name as well as their site. Massie's associate in the land business was Duncan McArthur, who acted as an agent for purchasers of Virginia military land warrants and worked as a surveyor for Massie, locating the lands. McArthur was paid in land for the warrants he surveyed, one-fifth of the acreage entered.[6]

In 1796 Thomas Worthington of Berkeley County came to the Scioto Valley and made a substantial entry of land with the assistance of McArthur and Massie. He returned to Virginia, married, and purchased more land warrants. In 1798 Worthington moved to Chillicothe, accompanied by his brother-in-law, Dr. Edward Tiffin. Massie, McArthur, Worthington, and Tiffin would emerge as leading citizens and politicians of the Virginia District. Among them, Massie was the greatest landholder, the owner of more than 78,000 acres in the Virginia District alone. McArthur became a military general. Worthington served as United States Senator and as governor of Ohio. Tiffin was president of the state Constitutional Convention and also was elected governor of Ohio.

They were strictly Republican: Ardent followers of Thomas Jefferson, they were opponents of the Federalists, including the first governor of the Northwest Territory, Arthur St. Clair. According to Andrew R. L. Cayton,

"Massie and his colleagues expected to be freely entrusted with power because they believed in the ability of the populace to recognize and reward obvious talents. Their material accomplishments made them worthy of office; popular elections simply confirmed their eminence."[7] Were these landed gentry of the Scioto Valley Upland Southerners?[8]

The Ordinance of 1787 prohibited slavery in the Northwest Territory, the antislavery clause itself being somewhat surprising at that time. Thus, no one who had slaves could bring them north of the Ohio River. Worthington freed his slaves in Virginia before moving to Ohio; they became "Negro servants" who moved with the family to Chillicothe in 1798.[9] Other Virginia families did the same, and soon Chillicothe had a small black population consisting of manumitted slaves. Worthington and most of his contemporaries were against slavery, though certainly not opposed to servitude; they commonly aligned themselves against measures that would promote racial equality.

Chillicothe was to the Virginia District what Lexington was to the Bluegrass. The fact that Ohio was "free" and Kentucky "slave" has tended to mask similarities between the two. Both were extensions of the Upland South that drew population from the Great Valley and the Piedmont. Neither had a landscape that suggested equality. The expansive country estates and manor houses of the Bluegrass had equally impressive counterparts around Chillicothe where the great landholders built themselves imposing residences on the hills overlooking the fertile bottomlands.[10] Worthington's estate, Adena, occupied a prehistoric burial mound atop the bluffs. In the Virginia District the early tracts of land were entered, as in Kentucky, in parcels of several hundred to a thousand acres each, following the Virginia manner of indiscriminate survey, sometimes called "zig-zag, cut, and carve," which allowed the surveyor to include or exclude any land he wished. Many tracts were held by absentee owners who rented the land to tenant farmers.[11]

Few towns were founded on the Scioto slope. Massie's town, Chillicothe, became the seat of Ohio government (territorial and state) and it performed all the urban functions of the area. Tenancy, land speculation, and large holdings depressed the rural population density and dampened the prospects for new towns in the hinterlands. As late as 1848 an agricultural correspondent from Madison County, in the heart of the Virginia District, wrote, "[t]hese lands . . . are owned in large tracts, which retards the improvement of the country, and tends also to prevent any great increase in population."[12] The District was, in these respects, less of a New Virginia than it was an extension of the old.

The land itself surely did not retard settlement (fig. 15). The meadows that Christopher Gist had described were instantly recognized for their great

Fig. 15. Intersection of Prairie Road and the Charleston-Chillicothe highway in the old Virginia Military District of Madison County, Ohio. These rolling prairies were an early cattle range that supplied the Scioto Valley feedlots.

value. Some tracts were open woodlands (savanna), others nearly unbroken grass. Grassy swards covered some of the river and creek bottoms. In Fayette County "the barrens" was a land "divested of undergrowth and tall timber . . . covered with grass." Approximately one-third of Madison County was in this category. Western Pickaway County was described as "prairie and white-oak plains." Marion County, at the northern tip of the Virginia District, was one-third prairie skirted by an open canopy of oak and hickory.[13] Much of the upland prairie was kept in grass to support herds of cattle, although a portion was plowed and put into oats; wet prairies were used to pasture cattle. The bottomlands, beginning along the Scioto River and Paint Creek at Chillicothe, were put into corn immediately.

If one had to identify a place where the Corn Belt begins, no better choice could be made than the broad plain of the Scioto at its junction with Paint Creek at Chillicothe. South of there the land is hilly, the bottomlands are not large, and agriculture is crowded by topography. The big cornfields start at Chillicothe today, just as they did 175 years ago. Today they stretch west from the Scioto, across the Miamis of Ohio, covering mile after mile: the Tipton Till Plain and the Wabash Valley of Indiana; the Grand Prairie and the Military Tract of Illinois; the loess plains and the Des Moines lobe

Fig. 16. Corn in the shock, Calhoun County, Illinois, 1988. The corn shock was a useful means of securing both corn and fodder in the field, a hallmark of cattle feeding in the system that employed open lots.

of Iowa; the Missouri Valley of Nebraska; and beyond there, up the irrigated Platte River Valley where today feed grains are loaded into railroad cars for shipment to the mouth of the Columbia River for export to Japan and Korea.

But even though the Corn Belt's westward projection may now be virtually completed, its origins trace even farther back east than the Scioto Valley. The feeder livestock industry that emerged in the Scioto Valley's cornfields in the first decade of the nineteenth century was the creation of a small group of men, most of whom came from the South Branch of the Potomac River in Hardy County, (West) Virginia, the second major valley west of the Shenandoah. The South Branch cattle feeders are credited with a number of innovations, including invention of the corn shock, a means of securing corn and fodder to feed cattle in open lots (fig. 16).

From the South Branch there were two migrations of cattlemen to the west, one to the Bluegrass of Kentucky, the other to the Scioto Valley. Felix and George Renick were prominent among those who went to Chillicothe. William Renick, George's son, later gave a good description of the method they employed:[14]

> On the Scioto river much of the land was owned by immigrants from the south branch of the Potomac River, Virginia, where the feeding of

cattle had been carried on for many years in a manner peculiar to that locality, and which materially differed from the mode practiced in Pennsylvania or further north. The cattle were not housed or sheltered, but simply fed twice a day in open lots of eight or ten or more acres each, with unhusked corn with the fodder, and followed by hogs to clean up the waste and offal.

George Renick scouted southern Ohio in 1797 with the intention of relocating there. Near Lancaster he saw a small prairie "growing in rank luxuriance," a sight that would attract any cattleman.[15] He moved to the Scioto Valley in 1802, located a prime piece of bottomland near Chillicothe, and was soon in the company of other South Branch feeders. In 1800 Scioto Valley residents had to import grain from Pittsburgh, but by 1803 there was a corn surplus. George Renick began fattening cattle at Chillicothe in 1804, and the next year he drove fat cattle from Ohio east across the Allegheny Mountains to the Philadelphia market, the first of many such drives he would make. His feat also demonstrated that doubts about the profitability of such drives could be laid aside.[16] By 1810 Felix Renick was importing stocker cattle from outside the Scioto Valley to supply the feedlots. At least five thousand cattle were being fed annually in the Scioto Valley by 1820. They wintered in open lots on corn and grass, were turned out on bluegrass pasture in the summer, and were fattened on corn to reach market weight before the annual drive the following winter.

Unpopulated grasslands of the interior counties in the Virginia District were the original open range, the source of stock for the Scioto feedlots. Demand kept increasing, however, and the westward reach for young cattle thereby grew. The Grand Prairie of Indiana and Illinois began supplying young cattle by the 1830s, and the prairies of Missouri were drawn in a few years later. The early decades of the nineteenth century thus saw the rise of the middle-western system of range-feedlot agriculture. Cattle were born on the western prairies, sold in the eastern markets, and fed to market weight on corn at an intermediate location. The Scioto Valley and the Bluegrass were the first of these intermediate locations that anchored the system.

River or canal transportation was occasionally used, but it was unimportant. Drovers and the herds they guided overland accomplished all of the transportation, on the hoof, until the days of the railroad. Immigrants heading west learned to anticipate meeting cattle drives. Traveling from Harpers Ferry in the fall of 1819, John Woods, who was bound for Illinois, "met two droves of fat beasts from the south branch of the Potomac river"; later his party encountered "a drove of 120 oxen from the State of Kentucky," large animals of mixed breeds bound for the Baltimore and Philadelphia markets.[17]

Ancillary operations of cattle feeding also provided income. One or two hogs could be fed on the solid waste produced by one steer that consumed sixty bushels of corn in the feeding season. Fields used for feeding one year were planted to corn the next, and that was usually followed in rotation by winter wheat. In the mid-1840s Pickaway County, Ohio, was producing more than two million bushels of corn annually, of which only a small fraction was sold for cash. The wheat exports of the county amounted to 25,000 bushels a year. Hogs were converted to bacon and pork (1.3 million pounds) or lard (1.5 million pounds). At least 7,000 head of fat cattle were being driven east to market annually from Pickaway County alone by that time. Other products, including flour, oats, and flaxseed, gave the county an annual agricultural income of about one million dollars.[18]

Kentucky's Bluegrass variant of the system was similar. The grass itself had arrived in Kentucky by some natural spread from the east, but mostly as seed carried in sacks by Virginians and Pennsylvanians who planted it in clearings or on old cane lands for the purpose of improving pastures. The Pennyroyal Plateau, already a grassland, developed as the range cattle portion of the state's industry, while cattle feeding was concentrated around Lexington, Winchester, and Paris. Eight Bluegrass counties (all small in area) harvested more than one million bushels of corn each in 1839 and, among them, fed or pastured more than one hundred thousand head of cattle (undifferentiated as to type in the 1840 census). Bourbon County sent ten thousand head of fat cattle to market in 1835 and produced forty thousand fat hogs; the rest of its corn crop went into whiskey valued at $70,000.[19] Other agricultural specialties of the Bluegrass, especially hemp and tobacco, were the dominant activities in some counties.

Bluegrass cattle feeders were friends as well as relatives of those in the Scioto Valley, but they maintained a friendly rivalry in stock breeding. Shorthorn cattle had been imported into the United States from Great Britain in 1783. Blooded cattle were raised on the South Branch of the Potomac by 1790 and then were driven west, first to Kentucky and from there to the Scioto Valley. Importations were made directly from England to both the Bluegrass and the Scioto Valley in the 1830s, and both introductions were successful. The beef cattle that would follow these importations fattened more rapidly and went to market heavier than was possible before. Foundation herds for cattle-feeding areas of the Corn Belt drew upon the improved cattle of the Bluegrass and the Scioto Valley, partly because the two areas were known for their cattle and even more because the business itself was taken west to the Mississippi and Missouri valleys by people emanating from there.[20]

Clover meadows in the Nashville Basin also were attractive to livestock men. A half dozen Middle Tennessee counties had more than twenty thousand cattle each in 1839. Pennyroyal grassland and Highland Rim woods served as the range for the Nashville Basin feeders. They were well positioned to drive cattle to several markets—northeast via the Great Valley to Pennsylvania or south to the plantation districts. But Tennessee's concentration was on pork production, not beef, and thus its large corn crop (largest of any state in 1839) is only partly explained by cattle feeding.

In the early decades of the nineteenth century fat cattle emerged mainly from the Scioto Valley and the Bluegrass. The fact that a single enterprise was established by individuals with similar backgrounds in areas so close together would not be remarkable except that Kentucky and Ohio are not thought of as a single region. In de Tocqueville's words, "the traveler who floats down the current of the Ohio . . . may be said to sail between liberty and servitude."[21] But Kentucky and Ohio shared much in common in terms of land and resources for the cattle business because of their similar histories of landscape modification. The grassy uplands of the Virginia Military District in Ohio and the Pennyroyal/Barrens of Kentucky were well placed to supply young cattle for the respective feeding areas. Because Virginia interests were projected into the Scioto Valley as well as into the Bluegrass, there were also similarities in terms of population origin. The "Northern" state of Ohio and the "Southern" state of Kentucky were both "the West" at that time.

Fattening cattle on shocked corn in open fields was a new development. Great herds of mixed cattle, including some Spanish breeds from Florida, had been driven north from the "Long Canes" and "Big Glades" of the Carolina Cowpens to Norfolk, Baltimore, and Philadelphia in the eighteenth century. Southern Anglo cattlemen eventually encountered Spanish cattle traditions in Texas and formed the range livestock industry of the Great Plains.[22] But what the South Branch feeders and their Scioto Valley-Bluegrass offshoots developed, instead, was a system that would suit the Corn Belt. And as it grew, the livestock industry of the entire Great Plains would supply cattle for feedlots across the Middle West, a massive projection of what existed in microcosm in the early nineteenth century Ohio Valley.

In 1832 William Renick built a new home for his wife on land that she owned overlooking the cornfields and feedlots of the Pickaway Plains between Circleville and Chillicothe. "Mount Oval" is said to have been based on plans made by Thomas Jefferson. Pillars, imported from Virginia, were incorporated into the gracious but unpretentious country house. Renick had so many cattle drovers as overnight guests that he included in the design a small bedroom for their use, off one of the porches but not directly accessible to the house.[23] Mount Oval, which still stands today, suited the practical man of some means that William Renick obviously was.

In his very useful history, *Cattle Kingdom in the Ohio Valley*, Robert Henlein suggested that the Scioto Valley feeders, like the Renicks, were "businessmen concerned with sending cash produce to a distant market."[24] They took risks, borrowed money, and expanded their operations in the usual manner of entrepreneurs in any business. Henlein also mentions that, since a market for corn was lacking in the Ohio Valley, "the solution was to convert the corn into marketable produce." His own evidence suggests that no such solution had to be sought, however. The South Branch feeders had already worked out the system of cattle feeding they would employ. They went west for the purpose of recreating in a better land the business they already knew worked well.

Standard explanations that farmers had to search for a viable system are not credible in this case. An earlier version was given by Bidwell and Falconer, who stated: "The early settlers of the Scioto Valley had found that although large crops of Indian corn could be raised with but little labor, there was no accessible and remunerative market for the crop. They had therefore devised the plan of fattening large herds of cattle on corn and then driving the cattle to eastern markets." Their account may have been inspired in part by William Renick himself, who once explained how the Ohio pioneers needed a source of cash; making flour was possible, although the only market was New Orleans, which was too difficult to reach, so they turned to cattle feeding.[25] But there is nothing about the behavior of his father, his uncle, his Kentucky cousins, or any others of record from the South Branch that would indicate that they went west without knowing what their purpose would be.

The precision with which the Scioto Valley cattle feeders organized their business is suggestive of mass production. A ten-acre cornfield was the basic unit, supporting one hundred head of cattle in the late fall fattening season. Twelve to sixteen hills of corn were gathered to provide one shock; one-half bushel of shock corn per head per day was fed from early November to February, when the drive to market began. For each field there were two or three of like size adjoining it, filled with the same arrangement of corn in the shock, through which the animals were changed to fresh lots at every feed, morning and evening. Each lot of one hundred head was fed together and then driven to market together.

Although farmers are sometimes said to take pride in their barns, this sentiment evidently was lacking in the Scioto Valley. Feedlots eliminated cattle barns. The considerable amount of corn that passes whole through a steer becomes unusable waste when produced by an animal fed in a barn stall, but in a feedlot that corn becomes accessible feed for hogs. Between one hundred and two hundred hogs cleaning up after corn-fed steers could be fattened simply on the waste. The manure produced both by cattle and hogs enriched the soil of the series of fields through which they were rotated in the feeding season, providing nutrients for the next year's crop.

This was an efficient system, to say the least. Labor requirements were fairly small per unit of output, and capital investment in buildings and machinery was kept low as well. The land required for cattle grazing during the summer months represented the largest outlay of capital other than supplying stock cattle, but this phase of the annual cycle was the easiest to externalize. Cheap land—often free land—in the West subsidized the cattle feeders by providing the great expanses of grass necessary to support grazing herds. Western drovers supplied a steady flow of young, grass-fed steers to the Ohio Valley feedlots. William Renick's claim in 1880 that this was "still the almost universal method throughout the West, having undergone but little or no change in fifty years," was more than a boast.[26] The same methods were taken to every cattle-feeding region of the Middle West. Not until irrigation developments in the Great Plains during the 1960s drew cattle feeding west to areas which, until then, had been only suppliers of stock cattle for eastern feeders, was there any noticeable change.

The South Branch feedlot, as elaborated more fully in the Scioto Valley, was a new mode of agriculture that had no obvious direct antecedent in any of the earlier agricultural districts of the eastern seaboard. That it was established in the Virginia Military District of Ohio by men who came from the western valleys of Virginia is beyond doubt. The District acquired some other hallmarks of Virginia or generally Southern heritage. "Shotgun" houses, narrow structures, one-room broad and several deep, set side by side, were common in plantation areas of the lower South, and some can be found today in older sections of Chillicothe. Named country houses like Adena, Mt. Oval, or Duncan McArthur's "Fruit Hill" also carry an aura of the plantation. Large tracts of land with irregular shape, old tenant farmhouses, court-house towns, and a countryside largely devoid of smaller trade-center towns (except those appearing later along railroads) are carryovers from the past that still can be sensed to varying degrees on the broad uplands between the Scioto and Little Miami rivers.

But to suggest a Virginia heritage is not to pinpoint the source of influence. This New Virginia that supplied the model for the cattle-feeding industry of the Corn Belt had been created not by planters from the Tidewater but rather by farmers who inhabited the valleys that would eventually form the border between the Old Dominion and the new nonslave state of West Virginia. The roots of the Virginia District thus trace to a politically divided section of the country, the Shenandoah Valley of Virginia and its parallel counterparts to the west.

The new state of West Virginia, populated by Upland Southerners, was created during the Civil War because the slave state of Virginia seceded from the Union. While it is true that West Virginia was admitted as a free state (the only conceivable condition under which a new state could have been admitted at the time), the antislavery margin, which included most of those in favor of a new state, was provided largely by counties oriented to the Ohio River Valley. The problem of extending the eastern boundary to include some of the better lands of the Ridge and Valley section, so as not to make West Virginia solely an Appalachian Plateau state, was complicated by the secessionist inclinations of those more eastern counties. One attempt to derail the statehood bill in Congress was a proposal to attach another thirteen Valley counties to West Virginia, which undoubtedly would have tipped the balance to make it dominantly proslavery, hence unacceptable. The loyalties of people living in all the counties that were to become West Virginia were divided about 60–40 in favor of the Union.[27]

The Virginia-West Virginia boundary that was eventually drawn had to be carefully interdigitated among the pockets of free and slave sentiments. Northern Shenandoah Valley counties raised troops to aid both the Confederate and the Union causes. There was substantial support for both sides of the conflict in Berkeley County also, from which many influential citizens of the Virginia District came. The counties of the South Branch of the Potomac, as well as the "Big Levels" of the Greenbrier, both of which contributed people and know-how to the Scioto Valley and Kentucky Bluegrass cattle feeding areas, were similarly split.

According to Henlein, "the men from the Big Levels and the South Branch were by outlook and culture lowland Southern." The Renick family had migrated from northern Ireland to Pennsylvania around 1720. By 1740 one branch of the family was living in Augusta County, Virginia, near Staunton, in the Shenandoah Valley. The other group of Renicks moved to Hardy County (then Hampshire) on the South Branch of the Potomac. No part of the Renick clan appears to have lived in what could be reasonably considered the Lowland South.[28] But neither had Thomas Worthington, of English Quaker ancestry via Pennsylvania, who was a slave owner and master of landholdings in Berkeley County.

The most detailed explorations of Upland South culture have focused almost exclusively on the material side of the trait complex.[29] Its spread is as well documented as any current in the stream of American cultural diffusion. Although the migration patterns that established the Bluegrass and Scioto Valley fit quite easily into the Upland South cultural diffusion model, there is reason to doubt the equation of migration patterns with the spread of only a single cultural system. Whether in questions of politics, scale of agriculture, role of tenancy, or even general outlook, many traits normally associated with just the *Upland* South are hard to find in the Scioto Valley. Any proslavery sentiments the early Virginians who came to Ohio may have had became a moot issue by the clause prohibiting slavery in the Ordinance of 1787, although the enactment of black codes and attempts at exclusion would linger in the politics of the Old Northwest for years thereafter.

Ohio stockmen later moved north and west, and so did their counterparts from the Bluegrass. Some of the best prairie land in the free states of Indiana, Illinois, and Iowa was settled from Ohio, while the slave state of Kentucky projected its agriculture west into the slave state of Missouri. Yet, just as slavery did not survive the move north of the Ohio River, neither did the grand style of country life and large holdings worked by tenants survive when cattlemen left the Virginia Military District and moved west into Indiana, Illinois, and Iowa. There were attempts to take these elements west, notably to the Grand Prairie, but they became recognizable outliers, not the common pattern. The practice of bringing in large numbers of western cattle to feed, raising huge corn crops, and undertaking long drives to markets remained, a direct legacy of the aggressive entrepreneurship first evident in the Ohio Valley.

Cattlemen, whether in the Virginia District or elsewhere, commonly have been accorded a status in history exceeding that which has been granted to the hog farmer. It is an old distinction, perhaps of ancient origin, but one that has persisted nonetheless. Even cattlemen who have lost money in the business are typically held in high regard by their fellow citizens. The hog, in contrast, has long been regarded as a practical animal, a "mortgage lifter," whose sale pays off the mortgage on the farm. As a recent USDA publication quotes one Iowa farmer, "hogs don't always carry the prestige of cattle, but you can't live on prestige."[30]

Important though cattle have been in the Corn Belt's history, however, they have never produced as much income as have hogs. While cattle have always fared well on the sweetest limestone soils or the lushest prairie grasslands, hogs have thrived in just about any area that offered something to eat. Partly for this reason it was the Virginia Military District's prairies that supported the bovine ruminant, while woodland clearings in the less favorably endowed Miami Valley simultaneously began specializing in hog production. And just as interest was focused on cattle breeding in the Scioto Valley, making a better hog became the common preoccupation of farmers in the Miami Valley.

6

Razorbacks and Poland-Chinas

EARLY NINETEENTH-CENTURY Ohio was a proving ground for agriculture that tested all of the farming types that would eventually become established in the mid-continent section of the United States. Since Ohio was first in the line of settlement, it contributed its own influence to the subsequent states of the Old Northwest and Great Plains. A regionalization of population origins and agricultural types emerged around the various land claims and purchases from Lake Erie to the Ohio Valley.[1] The public-land areas that lay between the early districts developed as mixtures and gradations between the types.

Ohio's northeastern corner was the Western Reserve for Connecticut soldiers of the Revolutionary War, a westward extension of Connecticut Valley agriculture, and a stronghold of New England/New York habits. The combination of wheat culture and dairy husbandry these Yankees brought to Ohio was later established in Michigan, Wisconsin, and Minnesota, although their style of farming would have little influence on the Corn Belt and, in time, would be modified by corn/livestock agriculture even in those more northern states. The other major region of northern Ohio, the Lake Maumee Plain (known as the Black Swamp in the nineteenth century) was wet and mostly forested.[2] The work needed to drain and clear it for agriculture would take many years, and thus northwestern Ohio does not rank among the regions of the state that influenced the agricultural frontier.

The dryland counties of north-central Ohio were settled by the Scotch-Irish and Germans from Pennsylvania after the War of 1812; those bordering Lake Erie were settled from New York about a decade later. Both New York and Pennsylvania produced large quantities of wheat, and its culture was taken to Ohio in the initial migration. Winter wheat, planted in the fall and harvested early the following summer, was the principal variety grown, despite the danger of winterkill, because diseases and insect infestations were less troublesome the earlier the crop was harvested. Ohio produced more wheat than any other state in 1839. The crop was hauled north by wagon or canal barge to Lake Erie ports, principally Toledo, Huron, Cleveland, and Sandusky, and then shipped east to Buffalo for milling.[3] Southern Ohio farmers also raised wheat, usually in rotation with corn, and flour from Ohio Valley mills was sent down the Mississippi to New Orleans on flatboats. But the state's supremacy in wheat was short-lived. By the 1850s northern Illinois and southern Wisconsin had become the new breadbasket of the nation as the wheat frontier pushed west.

Wheat sown on new ground, especially the better-drained upland soils, produced a good yield and as sure a cash return as any crop a farmer with market access might choose. Bottomland soils, built up over centuries of il-luviation, were richer, and they were often marked for initial plantings of corn, a more demanding crop. Some intended wheat farmers "reduced" their richest land several seasons with corn to lower the nutrient level before they would sow wheat.[4] Soil too rich in nitrogen produced heavy wheat straw, but little grain. Continued wheat monoculture soon reversed the nutrient balance to one of extreme deficiency, however, and fields became weed-choked and subject to insect invasions. Such episodes, common in all the northern states from New York to Minnesota, marked the passing of the wheat frontier. Those who were determined to produce wheat moved on.

Farmers who remained on the lands now too poor for either corn or wheat converted them to pastures, an option that was especially attractive to sheep grazing.[5] Sheep were found suited to the uplands of southeastern Ohio, where crop farming prospects had always been poorer. Wool producers also took over some of the former wheat lands in the northern counties. In 1850 Ohio succceeded New York as the leading wool-producing state; south-eastern Ohio produced more wool than any other region of the United States at that time. Sheep-raising would remain a specialty of Ohio, although it never became an important activity in the Corn Belt generally.

A few good valleys crease the Appalachian Plateau of southeastern Ohio, but they lack broad stretches of bottomland, and the area has no prairies. Corn/livestock specialists bypassed that section of the state initially and found little interest in rediscovering it later. The early infusion of New England into the Ohio Valley—the lands of the Ohio Company of Massachu-setts located north of Marietta on the Muskingum River—would have little influence on the Corn Belt, either in terms of agricultural practices or in numbers of later western migrants from the region. A break in land quality was the determining factor. Great crops of corn fed to cattle and hogs in the Scioto Valley by 1820 foreshadowed the Corn Belt system that would catapult west to Kansas by 1870. But in that same length of time, intensive corn/live-stock agriculture crept no more than fifty miles east of the Scioto River into the hills of southeastern Ohio.

The Scioto and Miami valleys were the two most important birthplaces of the Corn Belt. The regional label, "Miami Valley," refers to the area of southwestern Ohio drained by the Great Miami and Little Miami rivers, al-though adjacent southeastern Indiana is so similar that the whole region can be treated as one in terms of physical environment, agriculture, and settle-ment pattern. The meridional Ohio-Indiana border, begun at the mouth of the Great Miami River, approximately bisects a tributary area that grew around the great nineteenth-century river port of Cincinnati.

The lands of the Ohio Company of Massachusetts on the Muskingum and John Cleves Symmes's tract between the Great and Little Miami rivers are sometimes referred to as the Ohio Purchase and the Miami Purchase, thereby distinguishing their method of creation from the Ohio land grants made to Connecticut and Virginia, which were intended as bounty payments for Revolutionary War soldiers.[6] But the prices paid for the purchases were negligible in comparison with their value. Nor was it necessary to have an intermediate group of associates, as in the case of the Ohio Company, nor to have a proprietor, like Symmes, draw up a plan for settlement and then sell lands to individuals, although these lessons were yet to be learned in 1787.

Symmes presented his petition for Miami Valley lands to Congress less than a month after the Ohio Company of Massachusetts had received favorable treatment on their request, and to some extent he followed their lead. Rather than form a company of associates, however, Symmes chose the archaic instrument of proprietorship to control and administer his lands, apparently following the method of creation of the Jerseys, colonies with which he had become familiar after moving to northern New Jersey from Long Island as a young man in 1770.

To Symmes's credit was his recognition of the type of land that would appeal to the yeoman farmer, the sort of man from whom he had commanded respect as a leader during the Revolutionary War. An interesting feature of the Symmes purchase was the stipulation that one-seventh of the price of each land parcel could be satisfied with military certificates.[7] He promoted this aspect of his scheme especially in New Jersey, a state lacking its own "western reserve" but which would now be served partially by the Symmes tract. His appeal to New Jersey soldiers, as well as his general reputation in the state, created a special interest in the Miami Valley among the citizens of New Jersey who, like those in the other eastern states, were infected with the western fever of the times.

Philadelphia was the major market in the Middle States for all western lands, and this circumstance, plus the far larger available population of southeastern Pennsylvania compared with New Jersey, led to a substantial number of Pennsylvanians joining the move to the Miami Valley. Virginians and Marylanders from the Great Valley and disgruntled former Pennsylvanians—especially Quakers who had moved south to the Carolina Piedmont but who wished to live in a territory where slavery was forbidden—also were prominent among the early Miami Valley settlers.[8]

As sole proprietor Symmes set the purchase price of each piece of land, and he did so with a democratic inclination toward encouraging the settle-

Fig. 17. Upland farm in the Miami Valley, Warren County, Ohio. The Miami
Valley was more wooded and its surface is more rolling than are the better lands of
the Virginia Military District.

ment of farmers, simultaneously discouraging the entry of large land parcels
of the sort that were made in the Virginia Military District. This did not
eliminate land speculation in the Miami Valley, but it did give a character to
the landscape behind Cincinnati that was markedly different from the Scioto
Valley. The Virginia District consisted of rich bottomlands and broad expanses of upland prairie, whereas the Miami Valley was mixed woodland,
occasional bottoms, and scattered old-field clearings at the time of settlement.
The Virginia District was a mecca for cattlemen, but the Miami Valley held
little such appeal. General farming, with a mixture of crops and livestock,
characterized the early years of the Miami Valley, at the same time that large-
scale livestock feeding emerged near Chillicothe and Circleville on the Scioto.
The differences were not absolute, but the contrasts were evident and well
known at the time. The Miami Valley was the place for the small farmer to
get started, own his own farm, and practice agriculture on a scale to which
he was accustomed (fig. 17).

   In addition to its crops of corn and wheat the Miami Valley also grew
a great crop of small hamlets of the sort that dotted southeastern Pennsylvania and the better lands of Maryland and New Jersey. Some, with names
like West Chester, Red Lion, or Blue Ball, had an obvious Pennsylvania her-

itage. Larger town centers also were laid out, and along lines familiar to Midlanders. The proliferation of small, crossroads trade centers and the appearance of larger, county-seat towns reflected another sort of democracy on the land that had its polar opposite in the sparsely settled hinterlands of the Virginia District. The many small centers of economic and social congregation that focused life for the Miami Valley's well-settled majority of freeholders seem to have been a natural outgrowth of the benign system imposed by John Cleves Symmes.

The man himself had little idea how to create a town, however. As sole proprietor he naturally tried to promote a site for the great city that was bound to appear somewhere along the Ohio River between the two Miamis. His attempt at founding a city was North Bend, the closest feasible site near the mouth of the Great Miami and probably a good choice in 1789. But Symmes was bested by a group of speculators from Kentucky who laid out the town of Losantiville, a name formed from a reverse reading of syllables in "city opposite the mouth (*os*) of the Licking," a few months earlier.[9] Governor Arthur St. Clair rechristened the town Cincinnati, to suggest the soldier-turned-farmer image of the Northwest Territory, thereby earning the federalist governor the further enmity of Kentuckians. Chillicothe, a Southern city in the North, and Cincinnati, a Northern city within sight of slave territory on the south bank of the Ohio, became natural rivals.

The launching of Symmes's Miami Valley venture, the birth of Cincinnati, and the beginnings of the downriver trade with New Orleans all took place within a few years in the late 1780s. Settlers were eager to come to the Miami Valley, although the better lands that attracted them lay well back from the hilly margins of the Ohio River—that is, beyond the limit of Symmes's tract. The first petition Symmes made to Congress asked for two million acres; he reduced the request by half in 1788 and six years later settled for a patent on a little more than three hundred thousand acres. The northern limit of his lands still remained in doubt, however. Symmes had faith that his claims eventually would be substantiated, and he continued to sell lands to the north despite his lack of authority to do so. Finding they had paid for and made improvements on land Symmes had sold them illegally, the new Miami Valley farmers sought redress in the courts. Symmes fought the lawsuits, continuing apparently until he had exhausted his resources. He died in poverty in New Jersey in 1814.[10]

Beverley W. Bond, Jr., insisted that, despite the problems, Symmes's colony was a success. While it certainly was not a colony, and its naive administration broke the man who created it, the Miami Valley itself was a success even beyond what Symmes might have hoped. Success came when the Miami Valley farmers and Cincinnati merchants adopted the same strategy that their neighbors across the Ohio River were following, that of supplying the New

Orleans market. For sixty-five years, until 1855, when Cincinnati for the first time shipped more goods east than it did downriver, the Miami Valley and its port city thrived on their New Orleans connection with the rest of the world.

Investing in a great commercial entrepot on the banks of the Ohio River was something of an act of faith in 1789. The only feasible route of commerce between Cincinnati and the eastern states was the Mississippi River route via New Orleans, then under the influence of an often-hostile foreign power. River transportation was cheap as long as the movement followed the current downriver, but it was prohibitively expensive upriver and made even worse by the prospect of overland wagon drayage to the east beyond Pittsburgh. Downriver was the only way to go, and even the introduction of steamboats on the Ohio did little to reverse the direction of trade because of the difficulties of transport across Pennsylvania before railroads were built.

Shipments to New Orleans began in the spring of 1787, when Kentuckians, under the leadership of James Wilkinson, sent a boat loaded with hams, butter, and tobacco downriver and sold their goods to the Spanish in New Orleans at a handsome profit. In 1800, before the Louisiana Purchase included both banks of the Mississippi within the limits of the United States, some transplanted Massachusetts shipbuilders at the New England colony on the Muskingum River began building oceangoing vessels, which were sailed from Marietta down the Ohio and Mississippi to New Orleans and on to Cuba. The seagoing port of Marietta, Ohio, had its problems, but the very idea of constructing vessels of several hundred tons burden to ply the inland rivers as well as the Gulf of Mexico and Atlantic Ocean bespeaks determination to build long-distance trade on a large scale.[11] The growth of New Orleans as a river port paralleled the expansion of the Corn Belt as an agricultural region, the two growth curves remaining in close harmony until the 1850s, when railroads formed overland links to the Great Lakes and, soon thereafter, directly to cities of the eastern seaboard, eliminating the need to ship all bulky commodities down the Mississippi River.[12]

The circuitous New Orleans route from southern Ohio to Baltimore or New York was of no concern to the Scioto Valley cattle feeders, because their fattened steers walked to market. Hogs also walked, about ten miles per day like cattle, and thousands were sent east from Ohio "behind" cattle on the long drives. But hogs commanded a higher price, pound for pound, when sold as manufactured products than when marketed live. Freshly slaughtered beef, like pork, had to be sold immediately in the era before refrigeration made long-distance transport of dressed meat a viable option. Salting and pickling were the only means of preservation, although both of these methods of curing were ill-suited to beef compared with pork. The steer, which yielded no bacon, hams, or lard, thus walked directly to market, while the hog was transformed into its many well-known products and by-products

Fig. 18. The Miami Valley was the most important center of hog breeding in the United States. Farmers there experimented with various European lines for purposes of breeding an animal that would bring a higher price at the slaughterhouse.

and then shipped to market in barrels. This demanded cheap transportation, and the Ohio-Mississippi system provided it.

The Miami Valley's concentration on hog production resulted from its settlement by thousands of farmers familiar with—and more than a few expert in—the breeding and care of swine (fig. 18). The same men might have turned to other agricultural specialties they also knew, although none that were apparently at hand brought the return that hog production offered. By the mid-1820s, pork products headed the list of downriver shipments from Cincinnati. The first hogs were walked down to the Cincinnati packinghouses by farmers living only miles away from the riverfront. As the valley became more thickly settled, new bands of hog production formed concentrically around the old, until by the 1840s farmers were walking hogs to Cincinnati from Logan County, Ohio, more than one hundred miles to the north.[13] Many hogs raised that far up the valley were sold to farmers near Cincinnati who specialized in fattening hogs, and thus a two-stage system of range and feedlot, similar to that developed for fattening cattle in the Scioto Valley, also evolved for hogs.

The first packinghouse of record in Cincinnati was established in 1818, although smaller-scale farmer-packers operated before then. Indeed, 27,000 barrels of pork and bacon passed around the Falls of the Ohio at Louisville in 1810–1811; by 1820, the annual total had grown to 114,000 barrels. A pack of 85,000 hogs was reported for Cincinnati in the winter of 1832–1833. The spirit of the times was reflected in the often-quoted boast: "[i]t was Cincinnati which originated and perfected the system which packs fifteen bushels of corn into a pig and packs that pig into a barrel, and sends him over the mountains and over the ocean to feed mankind."[14]

Probably few who moved to the city in its early years anticipated the extent to which pork packing would dominate the economy. By the 1840s new packinghouses emerged in towns along the rivers and canals leading to Cincinnati, although the Queen City remained the pork-packing center of the Miami Valley (and of the entire United States until Chicago first took that distinction in 1862). Even the Miami Valley's hog production could not match the demands of the Cincinnati packers, and by mid-century many thousands of animals from Indiana and Kentucky were being added to the annual pack of half a million hogs reported at Cincinnati and Covington.[15]

Cincinnati entrepreneurs organized the business on this scale and made the city earn the nickname "Porkopolis." Probably more credit should be given to the farmers who continuously improved the breed of hog supplied to the packers. Having come west from Pennsylvania, New Jersey, Virginia, or Maryland with the swine they had kept as much for domestic consumption as for sale to commercial packers, farmers understandably began to search for new breeds that would fetch a higher price at the market. The animal that eventually resulted, which could be described as the generic Corn Belt hog, was of ancient ancestry, the product of the European wild boar, *Sus scrofa*, and the East Indian pig, *S. vittatus*.[16]

Hogs were unknown in pre-Columbian North America. Their history in the New World began with many separate introductions in the sixteenth century, the de Soto expedition being but one example. The natives' curiosity plus the animal's own wandering habit soon produced a dispersal among the villages and into the backwoods. De Soto's expedition had observed some Indian villages in the Southeast raising a small, barkless variety of dog for table food. But these natives abruptly turned away from dog meat once they dined on pork. Hogs are gregarious animals and they reproduce rapidly, usually two litters per year. They adapt well to woodland environments both rich and spare in level of sustenance. A "great plenty of Spanish hogs" were discovered fending for themselves on Bermuda in 1610, probably survivors of a past shipwreck.[17] By the middle of the eighteenth century, Spanish hogs were common throughout eastern North America.

Hogs thrived on the mast of roots, acorns, and beechnuts of the southeastern woodlands. They were ideal frontier animals and were kept in all of the seaboard colonies in a semidomesticated state, running untended in the woods for most of the year, then hunted out of their wild haunts when the weather turned cold and the work of butchering and curing meat could begin. Allowed to roam free, hogs typically revert in feral condition within a few generations to the body configuration of wild ancestors many generations removed, and it was this problem that eventually led to confinement of hogs so that the inevitable crossing with their backwoods cousins might be eliminated.[18]

The common "razorback," "shark," or "land-pike" hog of the Southern woodlands had the long snout, long legs, sharp back, and roaming disposition of his European ancestors and was equally at home when fed or left alone in the woods. But although the hog demanded little, his keeper eventually came to expect more, and thus farmers soon began to take more care in the confinement and feeding of these obliging creatures. By the early decades of the nineteenth century it became common to pen the hogs after they had feasted on the woodland mast in the fall and to continue their feeding by switching to corn, either on the ear, on the fodder, or ground as meal, on into the early winter months.[19]

Fattening hogs on corn increased their weight and, more importantly, it also improved the quality of the pork and lard. Hogs fattened only on woodland mast produced soft pork which was also oily and hard to preserve, whereas corn feeding made the flesh solid, the lard white and firm.[20] The practice of "hogging down" the mature fall corn became widely adopted, whereby fenced fields of unharvested corn were opened to hogs for feeding. Since one hog typically consumed about fifteen bushels of corn in the fattening season, before walking to market, there was naturally a keen interest in breeding hogs that increased their weight the most on the least corn, in the shortest time, and which would thus bring the greatest cash return on the investment. Razorbacks were large animals with strong legs, and they could be driven many miles to market, but they were slow to fatten and produced little meat or lard.

Interest in improving American swine was apparent as soon as the importation of English stock could resume following the Revolutionary War. The blending of European and Asian ancestries had already occurred in Europe, the common Berkshire hog from England being one example, and the various crosses were then imported to the United States. The greatest attention to swine breeding was concentrated in the lower Delaware River Valley and in Maryland. One of the first identifiable American breeds, the Chester County [Pennsylvania] hog, or Chester White, was the product of an 1818 importation from Bedfordshire. New Jersey farmers bred the Jersey Red, from which came the Duroc Jersey, known in both New Jersey and New York, which in turn was bred by the Clay family in the Bluegrass of Kentucky. The most common breeds brought to the Miami Valley were two English types, the Bedfordshire and Russia, and an American breed, the Byfield, which had originated in Massachusetts. Each possessed desirable traits: the Bedfordshires were good walkers, the Russias extra large, the Byfields good feeders.[21] But no hog was yet available that combined all three traits, despite the intense interest in improving breeds.

The turning point came in 1816 when the Union Village Shaker colony in Warren County, Ohio, brought from Pennsylvania a boar and three sows of another English breed, the Big China. Although the Shakers would later forswear the consumption of pork, it was they who introduced the best hog yet to arrive in the Miami Valley.[22] The Big China was a large, smooth animal; when crossed with the Russia and Byfield lines it produced a superior feeder. For two decades breeders' attentions were focused on this animal, the so-called Warren County hog, that walked to market and fattened rapidly. It became the principal breed taken west to the newly emerging hog-raising areas of Indiana after 1825.

There were two more significant importations of English animals to the Miami Valley between 1832 and 1839. The Berkshire was an extra-large hog which became instantly popular on account of its size, a reflection of the main objective of swine breeders at the time. But the Berkshires had poor feet and crooked hind legs; and, it was discovered, they required great care in feeding to fatten rapidly. The "Berkshire mania" of the 1838–1845 period was checked by the arrival of the Irish Grazier, a great traveling hog with extra length between the shoulder and ham.[23] Thus as the hog grew larger and its market weight increased, it became increasingly necessary to breed an animal that still traveled well but did not consume excessive amounts of corn.

The many hog breeds and their varying qualities and shortcomings also reflected the disparate requirements of Miami Valley and southeastern Indiana hog farmers. Those farming near the packing plants in the lower valley were concerned almost entirely with market weight; they wanted a hog that would gain weight as fast as possible on the minimum feed and care. Farmers farther north, who walked stock hogs down the valley to sell at the feedlots north of Cincinnati, needed an animal that traveled well. But that hog would bring a lower price at the feedlot than one that was known to gain weight rapidly. Thus the common desire in breeding was to perfect a single animal that would perform well in all these respects. And as corn/swine agriculture developed in its territorial extent as well as in its intensity, a keen competition emerged among many breeders, who had the same objective of achieving perfection in the animal that would become the standard of the industry.

The result was the Poland-China breed of swine, a cross of the Warren County hog with the Berkshire and Irish Grazier importations, made between 1840 and 1845 by one or possibly several farmers in Warren and Butler counties, Ohio, the heart of the Miami Valley. The Poland-China name was not official until 1872 when it was ratified by the National Swine Breeders convention in Indianapolis, and then not without controversy. The "Poland" portion of the name is obscure. The breed association's history claims the name was chosen "to designate the progeny of a particular animal that had been obtained from a farmer, Asher Asher, who was a Polander by birth and resided in Butler County," although this explanation seems unlikely.[24]

Two men figured prominently in the Poland-China's origin, and both illustrate the backgrounds of Miami Valley farmers. David M. Magie, who came to Butler County from New Jersey as a child in 1813, claimed to have originated the breed and gave it the name "Magie hog," by which title it was known in the Wabash Valley of Indiana and in Nebraska and Kansas.[25] John Harkrader, a native of Wythe County, in the southern portion of the Valley of Virginia, farmed in Warren County near the Shaker colony and claimed credit for breeding the Poland-China on animals sired by the colony's boar. Samuel Harkrader, John's brother, took the Poland-China west to Hancock County, Illinois, on the Mississippi River, in the 1850s. John, although reported "amiable to get along with," was killed in a brawl with another man who did not share his Copperhead views during the Civil War.[26]

No Poland-China pedigrees were written until 1876, by which time the qualities of the breed had long been known throughout the Corn Belt. Even by 1852 the Poland-China was becoming the standard of the Miami Valley. John Harkrader marketed hogs that year that weighed 400 lbs. at the age of eighteen months; later, this rate of fattening was improved to a weight of 300–400 lbs. at nine months. The pride in fattening animals was considerable, reflected by one farmer from Miami County, Ohio, who wrote: "The skill of our farmers has arrived at great perfection in the art of fattening the hog, enabling them to convert nearly the whole animal into lard."[27] While this was an exaggeration, the emphasis on lard, rather than on meat, continued to grow in the mid-nineteenth century.

Lard oil was replacing sperm oil for home lighting use. Pork lard contains about thirty percent stearic acid, a common long-chain fatty acid that is used as an ingredient in making candles, soaps, cosmetics, and lubricants. Kentucky produced large quantities of lard oil. Cincinnati had thirteen factories producing lard oil and stearin, with a total output of one hundred thousand gallons in 1843.[28] Cincinnati's pork-packing industry was thus also the foundation for its soap and cosmetics industries, although in the mid-nineteenth century the lard moved down the Mississippi River in barrels loaded on flatboats, along with the hams, bacon, and barreled brine pork. Cuba was a major importer of Miami Valley lard.

The advances in hog breeding made in the Miami Valley and the spectacular growth of Cincinnati's hog-packing and processing industries in the 1830s obscure the fact that Ohio ranked no higher than third in hog production during this period of rapid growth. Tennessee was the leading swine state until the Civil War, and Kentucky ranked second in total production; Indiana, Illinois, Iowa, and Missouri held the top ranks after that time. The amount of Cincinnati's output directly attributable to Kentucky is difficult to estimate, although the number of Kentucky hogs walking north to the Ohio River had to have been as large as the number coming down the Miami Valley in order to account for the production figures reported for Cincinnati and Covington.

Ohio farmers who boasted that the old, long-snouted "sharks" had been eliminated from their farms were well aware that the same could not be said for the states south of the Ohio River. The razorback was alive and well in the woodlands of Tennessee and Kentucky, multiplying rapidly, providing food for thousands all over the South. English swine were imported to the Bluegrass on several occasions, and both Sussex and Middlesex hogs were sold at auction at Paris, Kentucky. Breed improvements were made among Nashville Basin hog raisers as well.[29] But these two areas were not typical of their respective states. It was the lean razorback, not the fat Poland-China, that comprised the largest number of hogs in the backcountry of Kentucky and Tennessee, and thus statistics reported in numbers of animals are mis-leading if any attempt is made to translate numbers into pounds of produc-tion. Southern hogs went to market at a weight averaging 130 to 150 pounds at the same time as typical weights in Ohio were increasing to upwards of 200 pounds.[30]

The hog was preeminent in the Southern diet. As Sam Hilliard has noted, "Nowhere . . . have swine ever reached the importance in regional di-etary resources they did in the South during the three or four decades prior to the Civil War."[31] The possibility of Southern production shortfalls in the face of a strong regional demand for pork has been a matter of debate. Any Southern deficit was met by the Northern surplus, although a precise defini-tion of the balance rests as much on the designation of "North" and "South" as it does on agricultural production statistics. Eugene Genovese, noting the inferior quality of both Southern hogs and beef cattle, suggested that the South had to have been a net importer of these products. Imports resulted in part from the monocrop myopia of Southern cotton planters who were not interested in diversifying their operations to produce food.[32] Hilliard, in contrast, concluded that, despite pork deficits in some areas, the South as a whole managed to provide for itself reasonably well. If the South is defined as only the Atlantic seaboard and Gulf states, as Genovese apparently chose to delimit it, then there is no question that meat imports were large. Hil-liard's South, however, reasonably included Tennessee and thus the many thousands of hogs moving south from there annually became intra-regional trade in his scheme.

Hogs and cattle also were driven south from Kentucky during the antebellum period, beginning as early as 1792, when improvements were made to the Wilderness Road, the same route leading back through Cumberland Gap that had earlier served as the first entryway for settlers. From there, drovers bent north into Virginia to reach Eastern markets or continued south past Knoxville and then followed the French Broad River into North Carolina. An average of nearly fifty thousand hogs were annually driven south through Cumberland Gap between 1835 and the early 1850s. The French Broad route was used by hogs from Kentucky and saw even larger numbers from middle and eastern Tennessee in the 1840s.[33] Other drives of cattle and hogs headed south to the Tennessee Valley and the Coastal Plain of Alabama or to the Georgia Piedmont, all cotton-planting regions. At least part of the large corn crop raised in the valleys of East Tennessee in 1839 can be explained by the market for corn created by the nearly constant passage of droves of hogs moving south and east to trans-Appalachian destinations. The trade continued even after construction of the first rail links between the Ohio Valley and the Southeast. In 1858–1859 more than forty thousand hogs were shipped south from Chattanooga over the Western and Atlantic Railroad.[34]

Two distinct systems thus emerged in a single industry. In the Ohio Valley there was a growing number of pork packers who purchased hogs locally, prepared meat and various by-products for market, and shipped most of it downriver to New Orleans, from which point their cargoes were assigned to various domestic and foreign destinations. All navigable streams tributary to the Mississippi became potential locations for meat packing in this logistical scheme of long-distance shipments utilizing a cheap mode of transportation. The Nashville Basin and Pennyroyal Plateau were served by the Cumberland River, while the Bluegrass had the Kentucky River, both easy routes to the Ohio.

Eastern Kentucky and Tennessee, in contrast, had no such convenient river access to markets in the South—or anywhere else. In these landlocked Appalachian counties the solution was to walk hogs up the valleys, across the mountains, and then down to the livestock markets of the Piedmont and Coastal Plain. The principal divergence between the two systems was the role of corn-fattening before driving to market. As the outlines of the Corn Belt began to emerge, the advantage held by areas capable of producing large corn crops became more evident. The fraction of total hog production attributable to the backcountry shrank, because the razorback inevitably fared poorly in competition with the corn-fattened hogs produced on the better lands to the north and west. The razorback eventually disappeared and swine production has remained a Corn Belt specialty ever since.

7

The First Corn Belt

THE TWO DECADES centered on 1850 mark not only the emergence of the Corn Belt as an identifiable region, but also its rapid expansion north and west to the Mississippi and Missouri valleys. Two factors set the 1840–1860 period apart from preceding as well subsequent developments. This first Corn Belt was a cultural region, one that can be understood by tracing the migration patterns of people who originated in the five islands of the pre-1840 era. Corn/livestock farming spread fastest when it was taken to new lands as part of frontier migration, slowest when it involved a reeducation of established farmers already practicing some other mode of agriculture.

A second aspect of the 1840–1860 Corn Belt is its close association with river-based commerce, most significantly with the alignment of the Mississippi and Ohio rivers and their principal tributaries. Because the river system of the mid-continent has a north-south orientation, westward expansion during this period was also northward expansion up the Wabash, Illinois, and Mississippi valleys. The availability of good bottomland—part of it already cleared and ready for cultivation and much more easily cleared—was one factor guiding the path of expansion along the tributary valleys. The transportation access these same rivers provided reinforced the concentration of meat-packers at streamside locations.[1] There they had access to a dependable supply of animals driven in from the surrounding prairie and woodland farms, and access simultaneously to the rivers, which floated boatloads of manufactured goods down the Mississippi to New Orleans.

The Wabash Valley's inclusion in the Corn Belt by 1840 typifies developments of the time. That the Wabash would become an agricultural district surpassing even the Scioto and Miami valleys was hinted as early as 1750 when Christopher Gist was told by Indians and traders at the Twigtwee's town on the Miami River of the fine lands to be found as far west as the Obache. John Cleves Symmes briefly tried to promote a settlement along the Wabash near Vincennes in 1787, before he commenced his Miami Valley scheme. But the first European presence in the Wabash Valley was established by the French. The post they called Ouiatanon was built in 1717 on the north bank of the river about eighteen miles below the mouth of Tippecanoe Creek, south of present-day Lafayette, Indiana. Vincennes, which was the major French (later, British) post on the Wabash, was begun in 1732.

The Wabash was a strategic route for the French, offering a shorter and usually safer route of passage between the Great Lakes and Mississippi waterways than either the Illinois or the upper Ohio. The French route led from Lake Erie up the Maumee River and across the short portage to the Wabash near Fort Wayne, Indiana. The Wabash Valley was a French corridor until the displacement of the French from most of North America by the British in the 1760s. A route that saw the passage of many trading parties between Quebec and New Orleans, the Wabash was also the site of several trading posts and scattered habitations.

Eighteenth-century French settlers in the Wabash Valley practiced native agriculture but with the inclusion of some European kitchen-garden crops. When British topographer Thomas Hutchins visited Ouiatanon in 1762, he found a cluster of small villages housing about six hundred Indians and fifteen French families. Hutchins noted that the country "would produce all kinds of Grain Natural to the Climate were the Inhabitants to turn their minds to Cultivate the Lands which they seem Intirely to neglect, except some small Gardens and raising Indian Corn."[2] That Hutchins did not take the production of Indian corn itself to be evidence of "grain natural to the climate" is testimony to British attitudes toward maize at that time.

French and Indian settlements typically planted corn and relied upon it for their bread. *Habitants* at Vincennes reported harvesting 5,400 bushels of corn in 1767. Prairies, created by Indian burning, covered large portions of the broad alluvial terraces around the city, a setting very similar to the Pickaway Plains of Ohio. Alluvial terraces of the lower Wabash had supported corn crops for centuries before the French arrived. Nearly the entire course of the Wabash as far upstream as north-central Indiana was bordered on one or both sides by level plains. Some were inundated by seasonal floods; others, known as "second bottoms," were above flood height and hence were more attractive to settlers. Sequential occupancy of Vincennes's valuable lands continued. The French gradually left, but four hundred Americans were reported living there in 1787.[3]

The first Indian land title that Indiana territorial governor William Henry Harrison extinguished was the Vincennes Tract in 1803, an area that included part of the Wabash Valley and also the equally good land around the forks of the White River just to the east.[4] Harrison then turned his attention to the hilly fringes of southern Indiana, which he secured from the Indians in 1804–1805, a selection made more in strategic terms of opening a strip contiguous to the Ohio River than with an eye to land quality. In 1809 he took what would become the Indiana side of the Miami Valley corn/hog district and negotiated for another portion of the Wabash, around Terre Haute, accumulating lands in a semicircular fashion around the eastern, southern, and western edges of Indiana.

Throughout this period Harrison pitted will and wit against the great

Shawnee leader Tekamthi (Tecumseh), who along with his charismatic brother, Tenskwatawa ("the Prophet"), was adamantly opposed to further cessions and advocated total separation from the whites. In November 1811 Harrison led an expedition of U.S. troops and local militia up the Wabash, past the site of Ouiatanon, to confront the Indians. Their destination was the village at the mouth of Tippecanoe Creek known as Prophet's Town, named after Tenskwatawa, who had removed to that point from Ohio in 1808. Harrison's men arrived and camped in the great cornfields the Indians tended next to their village.[5] The Indians attacked early the next morning and the bloody battle ensued. Tecumseh's people eventually fled, but not before they had killed about one-fifth of Harrison's force and wounded many more.

Although the Battle of Tippecanoe settled no land questions, the War of 1812, which followed, marked an end to British intrigues among the Indians of the Ohio Valley. The settlement that John Cleves Symmes had planned for the Vincennes area in 1787 was envisioned as "a few picketed towns," as were most frontier outposts at that time.[6] But immediately after the War of 1812 the countryside became less menacing, and agricultural settlers from the East and South began to arrive in larger numbers. The communitarian Harmony Society purchased twenty-five thousand acres of land, left Pennsylvania, and settled a gently sloping plain along the lower Wabash in 1814–1815 (fig. 19). Between seven hundred and eight hundred people lived there before the colony was sold to Robert Owen (who renamed it New Harmony) in 1825.[7] The Harmonists were good farmers, and they set an example for other colonists who came to the lower Wabash Valley.

Even better known was the colony established by Englishmen Morris Birkbeck and George Flower a few miles upstream from New Harmony on the Illinois side of the Wabash in 1817. Birkbeck's colony was not particularly important in the larger scheme of regional settlement, but his own and other colonists' literate descriptions of the "English Prairie" have left us a better knowledge of this place than we have of any other on the prairie-forest farming frontier.[8] Neither Birkbeck's "Wanborough" nor Flower's adjacent settlement of Albion were great successes, although the latter at least survived and became the seat of Edwards County. Fewer than one hundred settlers ever lived on the acreage that Birkbeck and Flower purchased for their scheme. But the English settlement did offer a demonstration example of prairie pioneering, involving the colonization of at least the fringes of open grasslands by settlers mostly unfamiliar with such environments.

Morris Birkbeck was an enthusiastic promoter of prairie settlement, yet as Douglas R. McManis has pointed out, Birkbeck's own example was not a bold one.[9] He chose to use the prairie lands for stock grazing rather than crops. George Flower broke some prairie land and planted it to crops; other colonists followed his lead, built cabins in the woods, and planted crops of

Fig. 19. Log-pen structures, including a corn crib (left), at New Harmony in the lower Wabash Valley of Indiana.

wheat and Indian corn in the prairie openings. There seems to have been much experimentation with various cultures. Birkbeck, especially, was interested in planting hedges for fences.[10] Like many other isolated frontier colonies, their aim at self-sufficiency was sometimes high, other times low. At least once they were unable to feed themselves and imported wheat from New Harmony. Yet in 1822 flatboats loaded with corn, flour, beef, pork, and sausage produced at the English settlement were observed traveling downriver to New Orleans.[11]

Brian Birch estimated that a little less than one-fourth of Edwards County consisted of open prairie at the time it was surveyed. Much more was covered by open woodland or brush, leaving about one-half of the area in some type of forest.[12] Vegetation reconstructions like these, based on early surveyors' notes, are the best evidence of the pre-European-settlement plant cover we have. But they necessarily reflect the length of time the area lay unused between Indian and white settlement and thus also measure how long a time elapsed for the forest cover to reestablish in the absence of fire or clearing. Part of the great variation in land-cover types that was reported— including prairies, barrens, savannas, brush prairies, open woodlands, and groves—simply reflects seral stages in vegetation succession that began with variously timed human modifications of the landscape. Thus to label one tract "prairie," another "brush prairie," and a third "immature forest" is as much an identification of time in a succession as it is an assessment of natural qualities of the environment.

John Woods, who came to the English Prairie from England in 1819, referred to part of the area as "barrens," which he described as "nearly destitute of timber, but much overrun with scrubby underwood," surely an example of a once-prairie in the early stages of reverting to forest. But despite these many openings, settlers still engaged in clearing land. In his two years at the English Prairie, John Woods saw several "rolling frolics," a community activity in which "many trees are cut down, and into lengths, roll[ed] up together, so as to burn them."[13] The brushwood and roots on the trees also were added to the fire. It would seem that the rolling frolic was indeed an activity of land clearing (rather than logging mature trees for timber) and that the trees described were not large. It was a community effort to supply the greater labor required to clear land that had advanced beyond "barrens" in its stage of forest regrowth.

John M. Peck's *Gazetteer of Illinois*, published in 1834, listed nearly eight dozen settlements located on or at the fringes of Illinois's prairies.[14] Some were towns, but most were communities of dispersed farms inhabited by an average of thirty to fifty families each. Few were on the Grand Prairie, the large, continuous zone of grassland in east-central Illinois. In the mid-1830s prairie settlements still were concentrated on the uplands between wooded valleys in the southern half of the state. To what extent these many tracts were totally devoid of woody vegetation is not known, but the lack of heavy forest can be assumed in all cases to have been the condition at the time of white settlement and not the work of clearing by settlers. In contrast to the grassy or shrubby lands the early white settlers found along the major rivers, which were more likely the result of a combination of Indian clearing and burning (sometimes for agricultural purposes), Illinois's many upland prairies had seen little aboriginal agriculture and were the product of fires, which regularly eliminated woody plants from the smooth, well-drained surfaces. The map of prairie vegetation in Illinois is a map of smooth plains.[15]

Kentucky had been the most recent home of many prairie-border pioneers who moved north into Indiana, Illinois, and Missouri after the War of 1812. Some of them had been born in Kentucky, while many others had lived there but had been born in Virginia. These were the men and women whose family histories included migration westward along the Wilderness Road through Cumberland Gap; they were vanguards of the eighteenth-century trans-Appalachian frontier. Because Kentucky/Virginia was a heritage that many of these pioneers moving north of the Ohio River shared, it has been common to label them Upland Southerners.

McManis culled from various sources a description of their nature. They were "very poor farmers," he wrote. "Small patches of cropland produced food for the household members and were enclosed to protect the crops from marauding animals. The keeping of livestock was an easy adjunct to the hunting economy. Animals, chiefly hogs and a cow or two per family, received little, if any care." The typical Upland Southerner had only meager possessions. "Besides living in poverty, the Upland Southerners were generally illiterate. They read little and wrote less." They differed "from the southern 'poor whites' principally in that their isolation was physical rather than social."[16]

This is a fair catalogue of regional stereotypes, omitting only the usual reference to "highlanders" with their "Elizabethan flavor." And the comparison with "poor whites" likely would not have been made at the time, inasmuch as it was a somewhat later epithet, a "term of remarkable elasticity," according to C. Vann Woodward, "useful in identifying this troublesome class and at the same time disposing of the problem of accounting for them."[17] It would seem surprising, then, that such a class of people moved north of the Ohio River and, by the 1850s, managed to transform Illinois into one of the major agricultural states in the nation.

The extra ingredient necessary to achieve this success has generally been credited to the Yankees, who, although they arrived later than the Upland Southerners, were (again quoting McManis) "receptive to the idea of prairie farming." They had more capital and were better educated. They invented new implements (such as the steel plow) and, most important of all, were better farmers. It would thus appear that had it not been for the arrival of Yankees, the Corn Belt might not have appeared north of the Ohio River at all.

One way of understanding who these Upland Southerners were is to learn more about their origins. A map of birthplaces of the early migrants to Sangamon County, Illinois, can serve as a representative sample of the pattern (fig. 20).[18] Two hundred forty-five heads of families and single individuals who settled in Sangamon County before 1830 were predominantly born south of the Mason-Dixon line in the three decades following the American Revolution. (Abraham Lincoln, born in Harlan County, Kentucky, in 1809, arrived in Sangamon in 1831 and thus is not included.) The arrows on the map of Sangamon County birthplaces are only inferred, but the birthplace pattern is so common, so similar to many other counties in the same region, that it may be presumed that population movements roughly followed the routes shown.

Those born in Kentucky differed by only a generation from those born in Virginia or the Carolinas, as would be confirmed if the birthplaces of children in these pioneer families also were shown on the map. It was a single pattern of migration, one that began in southeastern Pennsylvania and was

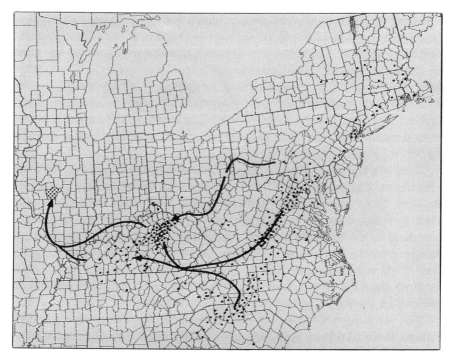

Fig. 20. Birthplaces of early Sangamon County, Illinois, settlers; see chap. 7, note 18. Migration routes, shown with arrows, are inferred.

supplied either from there or by new arrivals from Europe via the ports of Philadelphia or Baltimore. The first move was down the Great Valley (Shenandoah Valley in Virginia), beyond which some exited through Roanoke Gap and made farms on the Carolina Piedmont, a migration in full force by 1760.[19] Children born to parents who moved South returned north, joined by others from the Great Valley, to settle Kentucky in the 1780s, where the next generation was born. The overwhelmingly most important section of Kentucky involved in this phase of the migration was the Bluegrass region, the first (and at that time the only) settled section of the state. Given the variation in their ages, birthplaces of the Sangamon County immigrants reflect the entire sweep of their families' geographical history in America.

Two further generalizations can be made about the map. First, very few of those who settled Sangamon County were born in the plantation areas of Virginia or the Carolinas; they were "Upland" Southern in the sense that they were not from the Tidewater. But neither were more than a handful born in the true uplands, the Appalachian Mountains. Birthplaces are concentrated on the relatively good lands of the Great Valley, parts of the Piedmont, the Asheville Basin of North Carolina, the Bluegrass, and, to a lesser degree, the Pennyroyal Plateau. Thus Sangamon County's people were neither from the slave-oriented lowlands nor were they from the hardscrabble hill country; they were, in the main, farmers from the best agricultural districts south of the Ohio River or Pennsylvania. There were a few from New Jersey and New England, but as a whole Northern origins were unimportant.

How did these allegedly poor farmers fare in Illinois? In 1839, when the overwhelming majority of Sangamon County's farmers were Upland Southerners, the county produced 1.4 million bushels of corn, the largest crop in Illinois that year and comparable with counties in the Miami and Scioto valleys. Much of the crop was fed to the ninety thousand cattle and hogs kept on the county's farms. Rather than a land of subsistence, with small patches of crops, some hogs, and a cow or two on every farm, Sangamon County was developing corn/livestock agriculture on a large scale little more than a decade after it was settled. There was only a scattering of Yankee farmers in the county at that time, agriculture had already spread onto the prairies, and the county was exporting a substantial surplus. But this was no plateau in agricultural development. The 1.4 million bushels of corn represented 15.5 bushels per animal, not quite enough to include Sangamon County in the Corn Belt in 1840. Corn production nearly doubled by 1850 as more prairie land was broken—still without significant help from Yankees or any other regional population group except Upland Southerners.[20]

People of the same background that settled Sangamon County went to the lower Illinois Valley and to the lower Wabash and White valleys in Indiana. Sangamon County's map also resembles the origins of migrants to the Salt River Valley of Missouri, a large tract of prairie and mixed woodland southwest of Hannibal, which was settled mostly from the Bluegrass of Kentucky in the 1820s.[21] The major valleys and the better upland prairies of the new Corn Belt of Indiana, Illinois, and Missouri were populated by Upland Southerners, but not by poor farmers. The Bluegrass, Pennyroyal Plateau, and Nashville Basin, considered as a single region, constituted the most important source of migrants in the period before 1850. The Bluegrass and the Nashville Basin had well-developed agricultural economies; the Pennyroyal was less advanced, because of its later date of settlement. But it was a familiarity with good land, not poor, and well-organized agriculture, rather than backwoods subsistence, that the Upland Southerners moving north brought with them. As Morris Birkbeck himself somewhat enviously observed of these industrious farmers who had come up from the South, "[i]t is on the boundless scope for rearing and fattening hogs and cattle that the farmers place their chief reliance." In their statistical study of antebellum agriculture in the Old Northwest, Jeremy Atack and Fred Bateman concluded that the "Southern-born farmer was . . . no poorer than the Yankee-born farmer."[22]

Southwestern Ohio, including the Miami Valley and the Virginia District/Scioto Valley, was not an important source for these first migrations north of the Ohio River. Nor indeed could it have been, inasmuch as Ohio's settlement was later than Kentucky's and there was no large population of Ohio's sons and daughters ready to move west as early as 1815. Ohio was settled later, and hence its offspring moved to frontiers not yet accessible when the first outpouring of Kentuckians took place. Instead, it was the better lands of central and western Indiana and the Grand Prairie of eastern Illinois, opened to settlement a bit later but immediately west of the Miami Valley in the path of westward expansion, that grew out of the early Corn Belt islands north of the Ohio River.

Swine/maize agriculture spread rapidly across the rolling lands of south-eastern Indiana and reached the fertile, gently sloping Tipton Till Plain of central Indiana by 1830. With no need for acreages of grass for grazing, hog farmers were adapted to woodlands as well as prairie, although they needed to raise larger corn crops in a prairie setting where there was no woodland mast for their hogs. Miami Valley-style hog feeding thus spread incrementally to the west, leaving the dominantly woodland environment to move onto the prairies. By 1850 farmers were walking hogs to Cincinnati from farms around Indianapolis. North and west of there migrants from Ohio began to mingle with those who had moved up the Wabash Valley a decade or so earlier, and by 1850 the Corn Belt extended unbroken from the Scioto to the Wabash and beyond (fig. 3).

When cattlemen from the Virginia Military District looked west for new opportunities, they were not attracted to the wooded, rolling countryside of Indiana. They needed large acreages of grass, but little else. Few improvements to the land were necessary, and even poorly drained grasslands, unfit for raising crops, could be used for grazing. Not surprisingly, then, cattlemen from the Virginia District skipped over most of Indiana and established themselves on the large and then still uninhabited prairies west of the Wabash, especially the Grand Prairie of Illinois and extreme western Indiana. In time, these big-pasture cattlemen from Ohio began to intensify their operations after the fashion of the Scioto Valley, raising large corn crops to fatten their herds, but the initial migration to the Grand Prairie established extensive stock grazing rather than intensive stock feeding.

Benjamin Franklin Harris, who was born in the Shenandoah Valley in 1811, had teamed horses on wagons of goods moving across the Appalachians in the early 1830s. His father moved to the Virginia Military District in 1833, after which time young Harris became involved in cattle drives. He came west to the Illinois prairies to pick up a herd bound for the feedlots in Ohio and became eager to learn more about the business. In 1835 the cattle trail east from Illinois was a long one. Harris later recalled how they had to swim the Kaskaskia River at Vandalia, Illinois, the Wabash at Attica, Indiana, then drove through Muncie, Springfield (Ohio), and Columbus, where they swam the Scioto. Driving on to market required swimming the Ohio at Wheeling before crossing the mountains to Cumberland, Hagerstown, Gettysburg, and Harrisburg, finally reaching the point of sale, Lancaster, Pennsylvania.[23]

In 1841 Harris married Elizabeth Sage of Circleville, Ohio, and settled down on an acreage he had purchased on the upper reaches of the Sangamon River near present-day Champaign, Illinois. Others from Champaign County, Ohio, and its seat of Urbana, soon moved to the Grand Prairie in Illinois, where a new Champaign and Urbana were created. In 1855 Harris was raising 700 acres of corn, which he fed to 360 head of cattle and 200 hogs. Like other large landholders on the Grand Prairie, Harris rented part of his land to tenants, whom he relied upon to make some improvements and to thus effect the transition from range to feedlot. The Grand Prairie reflected its Virginia District heritage.

When B. F. Harris arrived in Illinois, corn/livestock agriculture was emerging in a cluster of lower Illinois Valley counties where floodplains along the river supported the first large corn crops. Corn was Illinois's staple crop by 1820, and by the 1830s both feeding and grazing had expanded onto the prairies.[24] By the 1840s, at least, the South Branch/Scioto Valley corn shock and feedlot were known in Illinois. While some farmers in Sangamon County cut their corn stalks, raked them into rows, and burned them, Amos Hilliard of Macoupin County wrote, "our best method is to cut up and shock the corn." He might as well have been describing a Scioto Valley feedlot when he wrote, "in the winter the corn is daily hauled into a lot, to be fed first to fatting cattle; which being turned out, stock cattle follow up to eat the fodder, and these by hogs." Joseph G. McCoy's description of activities around an Illinois feedlot of the 1860s similarly bears a close resemblance to William Renick's account of the Scioto Valley in the 1830s.[25]

The biggest Illinois cattle feeder at that time was Jacob Strawn, a Pennsylvania Quaker who had come to Morgan County (adjacent to Sangamon on the west) in 1831. Strawn fattened cattle, but used the Pennsylvania system of stall-feeding rather than the feedlot method. By 1854 Strawn was raising several thousand acres of corn on his ten thousand-acre homestead in Morgan County. Another Morgan County cattleman, John T. Alexander, a (West) Virginian who learned the cattle business in Ohio, came west with cattle in 1848 and was soon engaged in large-scale cattle feeding and shipping. Morgan County shipped seventy-five thousand head of hogs and cattle to the East in 1861.[26]

One of the best-known names in Grand Prairie agriculture is that of the Funk family, early pioneers in cattle and hog feeding who were later known for their work in the hybrid seed corn business.[27] The Funks, who came to the United States from the Rhenish Palatinate in 1733, moved from Pennsylvania south through the Great Valley, then to the Bluegrass of Kentucky, and in 1808 to the Scioto Valley of Ohio, yet another illustration of the typical migration pattern that established the Corn Belt. Isaac Funk moved west to what became known as Funks Grove, in McLean County, Illinois, in 1824. He raised cattle that he drove east to Ohio until 1841, and after that he began drives of cattle and hogs to the Illinois Valley packers. Eventually he drove stock across the 135 miles of prairie separating his farm from Chicago.

The Funks were among the first to make long drives to the city that would one day dominate the meat-packing industry of the Middle West. They drove two hundred to three hundred cattle and between eight hundred and twelve hundred hogs each trip, which typically required twelve days to reach Chicago. Hogs had to be fed along the way, and to supplement the corn they otherwise had to buy from local farmers, Funk's men started the trip with three wagons loaded with corn. By the time they reached Chicago the corn was gone and the wagons were used to carry hogs that had become disabled on the drive.[28]

Between 1840 and 1860 forty-five large-scale livestock enterprises, ranging in size from 920 to 26,500 acres, were begun in the Grand Prairie of east-central Illinois. Some were merely investments by absentee owners, but many were owned as well as supervised by men like B. F. Harris and Isaac Funk.[29] Among the absentees were William and Gustavus Foos, bankers from Springfield, in the Virginia District of Ohio, who put together more than 5,000 acres of federal and Illinois Central railroad grant lands in Champaign County, Illinois, between 1854 and 1865. They hired local managers, built tenant farmhouses, and leased land to tenants who would improve it. In 1877 the Foos holdings included about 1,500 acres of cultivated land, 400 cattle, and 160 hogs.[30] The slow rate at which the lands were improved reflected not only the mode of tenure but also the problem of drainage. At least one-third of the Grand Prairie was seasonally wet, the result of a hummocky topography produced by glaciation. It would require many years and substantial capital investment to bring all of it into crop production.[31]

Cattle were brought to the Grand Prairie of Indiana in the late 1820s. A steady trickle of new livestock entrepreneurs came in thereafter, and the area operated as a range for stockers supplied to Scioto Valley feedlots as well as a source of cattle for long drives directly to the Philadelphia market. Among the cattle barons of the Grand Prairie, none deserved the title more than Moses Fowler of Lafayette, Indiana. Fowler and his brother-in-law, Adams Earl, owned forty-five thousand acres in Benton, Warren, and White counties, Indiana, by the time of the Civil War. Their operations also included meat packing and a line of steamboats that plied the Wabash taking produce downriver to New Orleans.[32]

Fig. 21. Cattle grazed western Indiana's prairies before livestock feeding was established. The transition from grazing to feeding to breeding was part of agricultural intensification.

As in the Illinois Grand Prairie, the emphasis in Indiana was on scale; investments (except in land) were minimal. Occasionally, as many teams of oxen as could be found were yoked and put to work dragging a ditching plow through the heavy, wet, black sod in an attempt to drain the land of its seasonal inundation. But drainage efforts were more experimental than necessary to the mid-nineteenth century cattlemen. Edward Sumner of Vermont, Alexander Kent and Lemuel Milk from New York, and two dozen others, mostly from New York, were prominent among these "Hoosier Cattle Kings." The arrival of railroads and the growth of Chicago as a meat-packing center stimulated the cattle business and led to a more intensive agriculture on the Grand Prairie (fig. 21). Drainage for agriculture proceeded over the latter half of the nineteenth century to the point where Grand Prairie land eventually became too valuable to use only for pasture, although a mixture of land uses remained.[33] Grand Prairie breeders helped attract attention to Chicago as a center of purebred livestock. The largest single concentration of Hereford cattle imports in the United States was found on Grand Prairie farms within one hundred miles of Chicago.[34]

Paul Gates referred to these cattlemen as a "landed aristocracy of the prairies."[35] Noting their Yankee backgrounds (Daniel Webster once had an

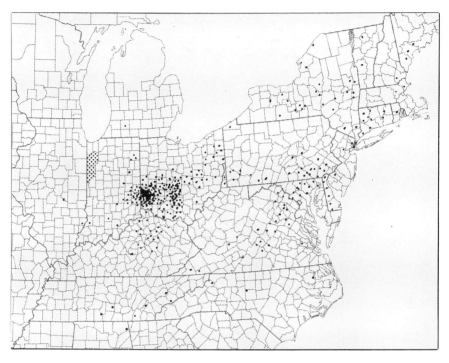

Fig. 22. Birthplaces of early settlers in Warren, Benton, Jasper, and Newton counties, Indiana; see chap. 7, note 36.

Indiana estate), Gates believed that the model for this development was the "cultured estates" of the Genesee Valley of New York. While it was New Yorkers who provided the capital used to assemble the large cattle acreages on Indiana's side of the Grand Prairie, New Yorkers did not dominate early migration to the area. Nor was the early style of farming traceable to Northeastern origins. Its obvious precedent, instead, was the Virginia District of Ohio, the same that influenced the Illinois portion of the Grand Prairie.

Warren, Benton, Jasper, and Newton counties, which were settled roughly in that order, cover most of the Indiana portion of the Grand Prairie. Biographical sketches of the four counties' first settlers confirm that the Miami Valley and the Virginia District/Scioto Valley were by far the most important sources of population (fig. 22).[36] The good alluvial lands along the Wabash River in Warren and Benton counties were settled first, and there the development was based more on corn feeding of hogs and cattle than on extensive grazing. Jasper and Newton, nearly all prairie land, had roughly the same birthplace pattern, though with the addition of more immigrants from the Northeast. Those who established agriculture in the Grand Prairie and the adjacent portion of the Wabash Valley in the early years were, in the main, farmers from the Miami Valley and Virginia District/Scioto Valley, the two Corn Belt islands north of the Ohio River. The connection is thus established much in the same terms as between these Ohio districts and the Grand Prairie of Illinois.[37]

Corn Belt agriculture expanded along the routes of river transportation until the 1850s. The Wabash, the Illinois, and the Mississippi were the major lineaments, and any area within fifty miles of a navigable stream was included in the river-based economy. Iowa was particularly well situated to participate in this system, and its marketed surplus helped St. Louis become the mid-continent's major entrepot. The subsistence-frontier phase was brief, if it existed at all. Iowa produced more than one million bushels of corn in 1839, only six years after settlement had begun.[38] The state's early farmers, who hailed from Ohio, Kentucky, Indiana, and Illinois, seem not even to have paused to consider any option other than corn-livestock agriculture. Early settlers occupied sites along the Mississippi and Des Moines rivers. As more immigrants from the same sources came to Iowa, the gently rolling loess-mantled uplands between the two rivers were occupied in similar fashion.

By the 1850s, and probably earlier, Iowa farmers were dealing in a well-recognized Corn Belt currency. Their large crops of Yellow Dent and White Gourdseed corn suggested the ancestral connection of Iowa to the Ohio Valley and, ultimately, to Virginia. New swine varieties included the Byfield, Irish Grazier, and China breeds, the same ones that were occupying the attention of farmers in the Miami Valley. A farmer near Keokuk reported in 1857, "most of the [corn is] husked on the stalk and placed in rail pens, there for the most part to be fed to swine and stock," the mention of "rail pens" being yet another indication of early Iowa's regional heritage.[39]

The Military Tract of western Illinois, an area between the Illinois and Mississippi rivers set aside as bounty lands for veterans of the War of 1812, exhibited Corn Belt agriculture only slightly later than the major valleys that bounded it. Upland Southerners dominated the early settlement of the Military Tract, although by 1850 settlers from north of the Ohio River numerically equaled those from the Southern states. The Sangamon country of Illinois also was a source of early settlers.[40] Those among them who were soldiers had come to redeem a promise made by the federal government: 160 acres of free land offered as an inducement to enlist at the time of the 1812 war. By 1816 the Pottawatomis had ceded all their land between the Illinois and Mississippi rivers north to the mouth of the Rock River, the location of present-day Rock Island, Illinois.

Although the lands were open to entry by military warrant holders, who presumably had no particular geographical concentration, settlement of the Illinois Military Tract was an extension of the same migration patterns that have been described for other parts of the Corn Belt: a northward movement from Kentucky and Tennessee and a westward movement from Ohio and Indiana. Scioto Valley-style feedlots were seen near Rock Island in 1857. An observer reported visiting an "extensive and prosperous livestock farm" where "two-hundred head of cattle and a like number of hogs were being fed." The place had only a single barn "and it was reserved for the horses." Another agricultural correspondent reported from Illinois that "barns are scarcely to be seen on the prairies, and they seem to be considered more of a luxury than a necessity."[41]

The relative absence of barns may be an indicator of Scioto Valley (rather than Pennsylvania) influence in the cattle-feeding business. When William Renick described the feedlot system invented on the South Branch of the Potomac, he contrasted it specifically with the system used in Pennsylvania. The economic advantage of feeding in lots was not only to eliminate the barn but also to allow hogs to feed on the substantial amount of corn that passed whole through a steer; barns prevented that. Pennsylvania, in contrast, probably gave birth to more types and styles of barns than did any other formative region in the nation's agricultural history.[42]

The feeder livestock barn, which eventually became common throughout the western Corn Belt, was not a Pennsylvania-type barn. It was basically an enlarged Upland South corn crib with shed wings attached (figs. 23, 24, 25). Hay was stored in the upper loft, while ear-corn was stored in cribs (pens) at ground level. Passageways through the structure, between the cribs or in the attached shed wings, were used to store wagons or implements and perhaps to shelter horses or young stock, but not to confine cattle for fattening. The feeder barn was a useful structure in a system of cattle feeding in which open lots were employed, and its eventual ubiquity in cattle-feeding areas is testimony to the widespread acceptance of the South Branch system.

Although Pennsylvania agriculture contained many of the elements present in the Corn Belt of the mid-nineteenth century, especially the interest in swine breeding and fattening meat animals on corn, there was not nearly the emphasis on these aspects in Pennsylvania that there was in the five Corn Belt islands. Pennsylvania farmers, as well as those who moved west from there, were still raising large wheat crops and, in general, practicing a comparatively diversified agriculture in the middle of the nineteenth century. In 1850 the northern limit of the Corn Belt lay roughly where Pennsylvania-stream settlers became numerically superior to those emanating from the five islands (fig. 26).[43] The basic difference was in the emphasis placed on wheat in this "extended Pennsylvania." Corn Belt farmers raised wheat only as a rotation crop, much as they produced oats. It took nearly a generation for the Pennsylvania-derived farmers of northern Indiana to adopt Corn Belt agriculture and thus reduce their reliance on wheat. In northern Illinois, where New Yorkers and Pennsylvanians predominated, it also took several decades to establish a large corn crop. The same is true for the northward spread of the Corn Belt in Iowa: farmers who began with a mixed system of wheat, corn, and livestock gradually learned to rely on the Corn Belt system that dominated southeastern Iowa.

The maps suggest that the Corn Belt, from its inception until the late 1850s, was a cultural region. It was born in the five islands of good land in Ohio, Kentucky, and Tennessee; and it spread into the new lands of the Wabash, Illinois, Mississippi, and lower Missouri valleys, all settled by people moving west from the five islands. Not only was there no significant Yankee influence in creating the region; neither did Pennsylvania itself, minus the offspring it sent to the Middle West via the five islands, directly shape the Corn Belt. Despite the Miami Valley's Pennsylvania-Jersey heritage it was more significantly the developments that took place in the Miami Valley, rather than modes imported from Pennsylvania and merely relayed through southwestern Ohio, that established the corn/livestock region of the Middle West.

Miami Valley entrepreneurs were also among the first to develop the marketing strategies that were necessary to establish Corn Belt agriculture on the scale it achieved in the era of river-based commerce. In 1803 merchants formed the Miami Exporting Company, a cooperative venture designed to better organize the downriver trade with New Orleans. Merchants had entered the business of milling grain and packing meat for shipment. They, as well as the farmers, needed a system that would handle the financial transactions involved in the three-cornered scheme of long-distance trade. In the Miami Valley, as elsewhere in frontier Ohio, merchants became manufacturers. Eventually, as the financial arrangements became more complex and the issue of credit more burdensome, the Miami Exporting Company evolved into Ohio's first bank.[44]

When corn/livestock agriculture moved west to the Wabash, Illinois, and Mississippi valleys, merchant-wholesalers entered the meat-packing business just as they had in the Miami Valley. As Margaret Walsh has written of these early packer-merchants, "in terms of commercial experience and ability to command cash or credit, they were the best placed personnel to undertake potentially lucrative but highly risky ventures."[45] Meat packing at that time was a purely seasonal business, one that was safe only after the weather had turned cold in late fall. Merchants who had an array of other activities to augment the annual cycle of business activities could engage in a sideline like meat-packing because it did not require year-around attention. They also had the warehouses necessary to store the product. And they were able to obtain the capital necessary to engage in long-distance transactions in which the farmer had to be paid for his animals, the work of packing completed, and shipment to market accomplished, all before any money returned to the point

Figs. 23–25. The basic corn crib evolved into the Corn Belt feeder livestock barn. The single crib was enlarged by attaching shed wings for storage (fig. 23, Grainger County, Tennessee) and then enclosed (fig. 24, Cooke County, Texas). The later Corn Belt barn (fig. 25, Cass County, Nebraska) remained primarily a structure for feeding, not sheltering, livestock.

of production. Wherever Corn Belt agriculture was found, there too were built slaughtering and packing operations that converted the animals into shippable commodities. Nearly all packers took a variety of stock brought to them, although pork packing was the major activity.

In the late 1840s Cincinnati accounted for one-fourth of all the hogs packed in the Ohio-Mississippi Valley (fig. 27).[46] Louisville, Madison (Indiana), and Chillicothe and Hamilton in Ohio accounted for another one-fifth. But outside those centers the industry was dispersed widely. Some packing houses were found in small cities away from navigable water, mostly serving only local demand. The Ohio Canal, completed in 1832 to link the Scioto Valley with Lake Erie at Cleveland, and the Miami and Erie Canal, which linked Toledo and Cincinnati after 1845, were attractive locations for meat packers as well, although the amount of the total pack that moved north, aimed eventually at New York via the Erie Canal, was comparatively small.[47] It was the Mississippi River and its major tributaries that attracted the largest number of pork packers.

By 1849–1850, packers were established from northwestern Missouri to

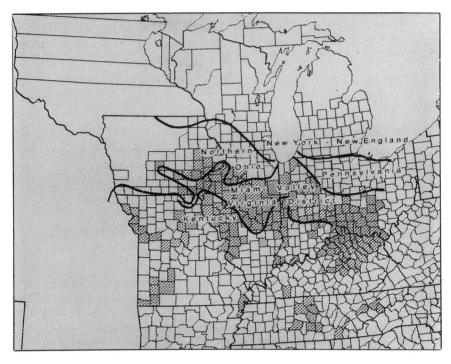

Fig. 26. Birthplace regions of middle-western settlers compared with the 1850 Corn Belt. At that time, Corn Belt agriculture was found primarily in the areas settled from Kentucky, the Miami Valley, or the Virginia Military District (see chap. 7, note 43).

the Cumberland River of Tennessee, their remarkable dispersion the product of Corn Belt agriculture's equally remarkable expansion. Farming had spread up the major valleys and then out onto the surrounding upland prairies. Any navigable stream with a population of hog-raising farmers within fifty, even one hundred, miles was a potential location for meat packing. While farmers made some shipments of corn and wheat, pork production brought the greatest dollar return. This was especially true for farmers some distance away from the river, who incurred substantial transportation costs hauling grain overland. Hogs walked the distance easily. At the same time, overland transport of barreled meat and lard was as high-priced as flatboat transportation was cheap. Thus the packers who were engaged in exporting products outside the region located their operations along the river, often right on its banks, and there received drove after drove of hogs walked in from the backcountry. A streamside location incidentally provided a means for disposing of the waste and offal, a deplorable practice only made worse in later years when

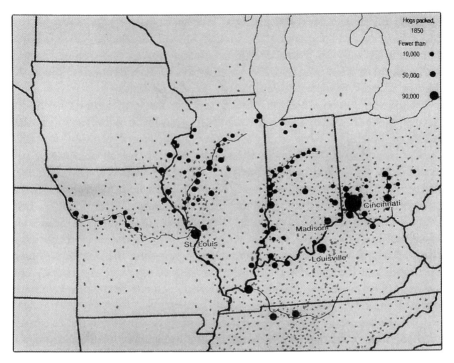

Fig. 27. The pork-packing industry was concentrated along the major rivers of the Corn Belt in 1850. Hog droving was the more important activity in landlocked areas of Tennessee and Kentucky. One dot represents 10,000 hogs reported in the 1850 census (see chap. 7, note 46).

packers lined the South Branch of the Chicago River (which later acquired the epithet "Bubbly Creek").

Outside the major centers, up the valleys, the typical packing operation in 1850 was small and had no local competitors; packing plants were spaced like beads on a string and the catchment areas of neighboring packers were likely discrete. Nearly the entire Wabash Valley was covered in this way, as were the Illinois, Mississippi, and Missouri valleys. In the Miami Valley, there were small-town packers who competed for animals with the larger, Cincinnati buyers. Canal transportation linked them with the Ohio River. Thus Cincinnati's large pack has to be explained in terms of a westward reach into Indiana and, perhaps even more, by southward market penetration into Kentucky, since a fair share of the Miami Valley's hogs had to have been purchased and processed before they could get to Cincinnati. The mid-size packing cities, like Madison and Louisville, carved out markets within the larger system. Packing plants existed along the Cumberland River, in the

Nashville Basin, and on the Pennyroyal Plateau, at least by the 1840s. By this time, the drives of hogs along the French Broad into North Carolina could have been fed only by a smaller hog-producing region, one confined largely to Appalachian Kentucky and the good valleys of East Tennessee, given the amount of pork reported coming down the Cumberland.

Migration patterns had established a single mode of agriculture in the Mississippi-Ohio Valley that was based on widespread adoption of similar breeds of hogs and cattle and the common practice of planting Southern Dent corn and using it to fatten the stock. The fact that all of this area was accessible to New Orleans via downriver shipments helped standardize marketing and packing procedures throughout the Corn Belt. Shipments of pork and pork products to New Orleans therefore cannot be assumed to have originated north of the Ohio River, inasmuch as the major pork-producing regions of both Tennessee and Kentucky were fully as tributary to the Mississippi River as were Illinois or Indiana. Even adjusting the numbers of hogs to compensate for the lower slaughter weights of some of Kentucky's and Tennessee's contribution, the downriver trade from the two Corn Belt states south of the Ohio River was substantial.

This mid-century geography of meat packing was rather short-lived, not from any lack of success in the business, but rather because the pattern was typical of mercantile arrangements in a river-based frontier economy. The construction of railroads had only begun in the Middle West in 1850, but by 1860 railroads had so changed the pattern of commodity movement that the river-based system was beginning to fade.[48] The rivers flow from north to south; railroads crossed them, running east and west. The Corn Belt's matrix of trade flows was rotated about ninety degrees to the east, and traffic began heading toward Chicago rather than to dozens of isolated packing centers along the south-flowing rivers.

Railroads built to haul raw materials east and finished goods west also carried people, in all directions. By the late 1870s it would no longer be possible to speak of the Middle West in terms of population origins based on early settlement by one regional culture rather than another. Railroads effectively homogenized migration flows, making any new western territory within their reach almost equally accessible to any of the longer-settled areas. Efforts by railroad companies and various states and territories to induce migration and foster settlement produced colonies on the western plains that ignored the old westward-expanding regional frontiers. Thousands of European immigrants who settled the fringes of the Middle West by the 1870s freshened the existing regional patterns.

Even though the association between population origins and corn/livestock agriculture began to fade around the region's margins, that did not mean that the Corn Belt as a whole lost its earlier identity. Prior to the Civil

War all of the counties that made up the Corn Belt were dominated by people who had been born in or at least lived in the five islands of good land west of the Appalachians. Yankees (New York or New England ancestry) they were not. The Great Valley of Virginia was their most common historic bond, a land of small farmers when most of their ancestors had passed through there, but which, by 1860, had become integrated into the slave economy of the South.

At the beginning of the war the Corn Belt spread over seven states: Ohio, Indiana, Illinois, and Iowa (free); and Tennessee, Kentucky, and Missouri (slave). Kansas and Nebraska, their easternmost fringes now entering the Corn Belt, had been opened to settlement following repeal of the Missouri Compromise and a reopening of the slavery question in the West. Despite a roughly shared regional ancestry and a common orientation to corn-livestock agriculture as a means of livelihood, the Corn Belt as a whole did not form a region of political homogeneity in the mid-nineteenth century. Nor did the region, as such, respond to the war unequivocally. Sectional divisions had already been drawn through a sorting effect in the migration process that sent some Southerners to free territory, others to slave. That circumstance, rather than a single Corn Belt response, would emerge as the decisive factor in the sectional geography of that period.

8

Corn Belt Sectionalism

BEGINNING WITH FREDERICK Jackson Turner and his students there has been a tendency to simplify middle-western population origins into two American-born categories, Upland Southerners and Yankees. Both have been the subject of caricature, especially the Southerner, who rarely has been taken seriously. Both have been treated ahistorically even by historians who, with a wink and a nod, have romanticized them with tongue-in-cheek descriptions sufficient to call up the traits they supposedly possessed. Upland Southerners rarely have been damned, but they seem to have absorbed a disproportionate amount of faint praise. "Honest" is one such word, applied unstintingly by those searching for something good to say about them. As Malcolm Rohrbaugh has written, Upland Southerners were "poor but honest folk."[1]

The gratuitous "honest" serves two purposes. One is to distinguish the character of the rude, Upland South backwoodsman from that of the educated, get-ahead Yankee in the pageant of middle-western culture, a drama that pits two separate and unequal classes against one another in a struggle for regional dominance. Even Richard Lyle Power's *Planting Corn Belt Culture*, which made no apology for the contribution from the South, gave Yankees the high ground. Power claimed only that "the Yankees . . . under-calculated the taciturn latent strength of the upland culture." But he went on to add, in any case, "the Yankees who went to the West were not all from the . . . purest of New England stock."[2] William Vipond Pooley thought that in Illinois the Southern settler had given way "to the more energetic northern people," possibly because of the Northerner's "faculty for adapting himself to his environments." He was thus "destined to succeed in the conquest of the prairies."[3] According to Henry Clyde Hubbart, Southerners in the Middle West "blunted the edges of angularity which settlers from New England and New York revealed, and thus helped to give the later middle westerner certain affable southern traits." It was a confrontation between "zestful, rollicking Kentuckians and Virginians" and the "frugal, thrifty, curious, and austere New Englander."[4]

The first act of the pageant takes place on the eastern seaboard, where the opposing cultures gather their strength. The second opens with a march west over the Appalachians to the testing ground, the Old Northwest; Southerners arrive first, Yankees a bit later. The script for the third act has been much harder to write:[5]

The Scotch-Irish southerner lived in a warm world of family ties and emotional coloration, centered on his self-contained homestead. But the enterprising Yankee, to whom "a friend is more valuable than a relative," envisioned a rational world of contractual relationships, in which the farmer became a unit of specialized production in the widening web of a commercial society. The opposing styles would interpenetrate in a thousand subtle ways as the Midwestern character gradually jelled.

The finale, then, is not a battlefield victory for either side so much as, perhaps, a sort of regional melting pot.

A second reason for the faint praise of Upland Southerners has been the need to spare them from the damnation reserved for their Lowland cousins. Slavery, of course, is the issue. Since the Scotch-Irish hillbilly had no more use for slaves than did the abolitionist Yankee, the absence of slavery in the Middle West can be explained by emphasizing *Upland* in the region's Southern population. The problem with this explanation, as previous chapters have shown, is that the Southerners who helped create the early Corn Belt were not mountaineers from the Appalachians. They were, instead, farmers from the best agricultural lands of the Ridge and Valley and Piedmont as well as their offspring born in the islands of good land immediately west of the Appalachians.

Slavery haunted their old homelands. By 1820 the Bluegrass, Nashville Basin, and Pennyroyal Plateau were, apart from the lower Mississippi Valley and the Tennessee Valley of north Alabama, the heaviest slave concentrations west of the Appalachians.[6] Early Kentuckians brought slaves with them from Virginia, a practice repeated in the 1810s when Kentuckians moved to Missouri. By 1860, a belt of slave-holding counties stretched unbroken, east to west, across the state of Missouri. Slave-oriented Corn Belt counties in Tennessee, Kentucky, and Missouri were fringed by others practicing the same type of agriculture but where slaves were less numerous. As in the Cotton South, the largest numbers of slaves were concentrated on the best land. Since the Corn Belt system of agriculture typically occupied the best land to be found in the states where it spread, it is perhaps not surprising that slavery was found in those Corn Belt states where the practice was not severely limited.

While slavery might be expected in Kentucky and Tennessee, it somehow seems anomalous in Missouri, which is as far north as Illinois, Indiana, or Ohio. The Ordinance of 1787 excluded slavery from the Northwest Territory, roughly, the states north of the Ohio River and east of the Mississippi. But Missouri became part of the United States in the Louisiana Purchase of 1803; and while territorial governments were later established along lines similar to those in the Old Northwest, there was no prohibition against slavery in the

Louisiana Purchase. At the time of statehood in 1821 there were approximately ten thousand slaves in Missouri, scattered along the alluvial bottomlands of the Mississippi River (where the French had kept slaves earlier) and, increasingly, in the better uplands back of the Missouri River in the west-central portion of the state. The Missouri Compromise of 1820 led to Missouri's admission to the Union as a slave state and proclaimed its principal boundary with Arkansas, latitude 36° 30′, as the slave/free border for the admission of future states in the Louisiana Purchase.

Slavery was illegal in the territory out of which Ohio, Indiana, and Illinois were created, but that did not mean there was no support for it when new constitutions were being drafted. A substantial majority of the people were Southerners simply by circumstances of settlement history. Petitions asking to bring slaves into the Virginia Military District were presented to Ohio's first territorial legislature in 1799 but were refused on grounds that the Ordinance of 1787 forbade it. A proposal to the state constitutional convention of 1802 for a limited form of slavery in Ohio was defeated, 5–4, in committee, with Ephraim Cutler, a dominant figure in the New England settlement at Marietta, casting the deciding vote against. The proposal forbidding slavery then passed in the entire constitutional convention by a margin of only one vote.[7]

In Indiana, territorial Governor William Henry Harrison, son of a planter from the James River of Virginia, professed to believe that a majority in Indiana favored slavery. A constitutional convention at Vincennes in 1802 produced a proslavery vote, although the U.S. Congress took no action on the measure, perhaps because it violated the 1787 Ordinance. Undaunted, Harrison and the territorial judges then borrowed a Virginia law providing that slaves brought into the territory would perform service due their masters. In 1805, the proslavery majority in the Indiana General Assembly succeeded in passing a bill allowing slaves purchased outside Indiana to be brought in and bound into service.[8] Despite the proslavery sentiments, there were likely never more than 250 slaves in Indiana, although federal censuses continued to find a few slaves in the state as late as 1840.

Illinois's state constitution of 1818 forbade bringing slaves into the state thereafter (except on a limited basis to operate the salt works on Saline Creek), but it did not abolish existing terms of slavery and indenture. The 1820 census recorded 917 slaves in Illinois, the most of any northwestern state, although less than one-tenth the number in Missouri Territory that year. Nonetheless, Illinois had about 750 more slaves in 1820 than it had counted in 1810, and without the constitutional provision the trend might well have continued.

Seeking to amend the constitution, proslavery Illinoisians secured a bare two-thirds majority in both houses of the Illinois legislature in 1822, the necessary margin required to submit the question of a constitutional convention to the people. In the popular vote, taken in August, 1824, the majority voted against a convention, 6,640 votes to 4,972 in favor.[9] Nine southern Illinois counties were strongly proconvention; another half-dozen were about evenly split. The strongest margin against calling a convention that would have re-opened the slavery question was provided by the future Corn Belt: the Wabash Valley, the Sangamon country, the lower Illinois Valley, and the good uplands east of St. Louis. Counties where the proconvention majority was largest were among the hilliest and had the least productive soils in the state, those where slavery would have been a doubtful strategy given the agricultural possibilities at hand. While the people of all of Illinois's counties in the mid-1820s came from regional backgrounds focused on Kentucky, Tennessee, and Virginia, there had already emerged a geographical variation in pro- and antislavery sentiments within the state, an early sign that economic and social factors were vying with sectional feelings.

Although the slavery issue can be portrayed in simple terms as an opposition between "slavers" and "abolitionists," there were at least three distinct positions on the issue that should be taken into account in order to understand the events that would transpire by the 1850s. There were radical abolitionists who opposed slavery on moral grounds, doing everything in their power to check its further extension and, ultimately, to eliminate it from the land. They stood in sharp contrast to the proslavery citizenry, who contrived various economic and social justifications for its continuation and growth. In between those two extremes were many others who advanced more qualified arguments. Slavery was opposed by many in Ohio, Indiana, and Illinois because they wanted no African-Americans in their midst; slavery should be kept south of the Ohio River and their states ought not to serve as refuges for runaways or even freed slaves from the South. This third position is the one that most characterized the Corn Belt.

Despite the several attempts to broaden the scope of slavery in the Old Northwest, to establish the practice after it had been forbidden, there was little danger that slavery would have become widespread.[10] Support was qualified and opposition grew steadily as more settlers from Northern states entered the Ohio and Mississippi valleys. That is not to say, however, that the politics of race revolved solely around the issue of slavery. The so-called "black laws" were more popular than slavery was unpopular. An 1807 Ohio law required persons of color to post $500 bond to live in the state; township overseers were to remove paupers who did not oblige. The law apparently was used to forcibly eject blacks from Cincinnati in 1829. Blacks coming into Illinois were required to file a certificate of freedom, a law enacted in 1819 that also required them to present passes when traveling.[11]

Such laws were aimed not at slaves but rather at any person of color who happened to be in the state. Black laws had the support of slavery advocates but, more significantly in terms of numbers, they also enjoyed popularity among the antislavery majority, which sought to discourage the immigration of blacks. By 1840 the laws were an issue of contention between Democrats and Whigs, although neither party as such can be identified as instrumental in efforts to repeal them.[12]

Two arguments have been made as to why slavery never became accepted in the Corn Belt. The issue of moral repugnance to the institution itself was a contributing factor but alone cannot account for the outcome. The early settlers of southern Ohio, Indiana, and Illinois were divided in their opinions, and proslavery minorities mobilized several times to challenge the issue. Some Southerners, watching the passage of slaves across Illinois as they were taken to Missouri, argued that the state was losing the prosperous class of slave-owning farmers.[13] Attempts to allow a modified form of slavery suggest a divided populace, while the narrowness of the margin in favor of the status quo shows there was no overwhelming feeling of repugnance toward slavery in Illinois or Indiana at that time.

The other argument claims that slave labor was uneconomical. Carville Earle has suggested that the reverse held true by the 1850s—slaves probably were cheaper than hired labor in the lower Middle West by that time.[14] Economic models which show that one form of labor is cheaper than another rest inevitably on estimates of the costs and productivity of each. Such models can be used to justify an advantage for either free or slave labor, depending upon the assumptions. While it is impossible to prove or disprove that slavery induced or retarded economic growth in any state, a need for slavery—equally, a need to abolish it—became difficult to demonstrate in the Corn Belt as agriculture developed, apparently equally well, under conditions of free labor or slave. The lack of a clear answer to the labor question emphasizes that radically different social systems could exist simultaneously in different parts of a single agricultural region: slaves in Tennessee, Kentucky, and Missouri; free labor in Ohio, Indiana, Illinois, and Iowa.

If there were no convincing economic arguments against the profitability of slave labor, and if moral revulsion against slavery was less apparent in the Corn Belt than it was farther to the north, it is worth questioning why the region was not more apparently proslavery, especially given the Southern backgrounds of many of its people. Except for Illinois in 1820, where slaves outnumbered free coloreds 917 to 457, the free colored population was larger than the slave population in Ohio, Indiana, and Illinois in every census through 1860. Illinois's 1820 entry may be explained by its location across the path of migration from Kentucky to Missouri; about half of its slaves were found in the bottomlands of Randolph, St. Clair, and Madison counties, fringing the Mississippi River and opposite Missouri.

It appears that slavery was already an anachronism in Illinois and Indiana by the time Anglo-American settlement commenced. With a few notable exceptions, slave-owning continued in the nineteenth century at isolated locations where it had been established by the eighteenth-century French. More than half of Indiana's slaves in 1820 were held at Vincennes and the nearby forks of the White River, where the French had kept slaves in the 1760s. The concentration of Illinois slaves along the Mississippi between Kaskaskia and the American Bottom similarly reflected the persistence of a pattern established by the French. One of the terms of Virginia's cession of its western lands in 1783 had been the stipulation that property and lands of the French Canadians around Kaskaskia be recognized, a circumstance which virtually legislated slavery into existence there.[15] Kaskaskia and its vicinity held about one-fourth of Illinois's slaves in 1820 and 1830, the largest single concentration in the Old Northwest.

Urban or industrial clusters accounted for most of the slaves away from the old French strongholds. The second largest concentration in Illinois was at the salt works inland from Shawneetown. Slaves were used in the Missouri lead mines, and they were similarly employed by the lead industry around Galena in northwestern Illinois; Jo Daviess County reported thirty-one slaves in 1830. Nearly all of the rest of the slaves in Illinois and Indiana were found in settlements fringing the Ohio River. One concentration appeared for a time in Clark's Grant, the 150,000-acre tract opposite Louisville reserved for the officers and men who had marched with George Rogers Clark. There were eighty-one slaves in Clark's Grant in 1810, but the number was reduced to zero by 1820.

Although Illinois counted more slaves than would have been the case were it not on the route to Missouri from Kentucky, some Illinois slaveholders had left for Missouri after the 1787 Ordinance was proclaimed. Missouri's portion of territorial Louisiana's population was nearly 15% slave in 1810, before the immigration of slaveholders from Kentucky and Tennessee began in earnest. The old slaveholding pattern of Missouri, based on the strip of French settlements along the Mississippi and the lead mines west of Ste. Genevieve, was soon replaced by developments that had no counterparts in Illinois or Indiana.

Immigrants to Missouri brought thousands of slaves with them. By 1830 the older French counties, including St. Louis, could account for only about one-fourth of the slaves in the state. Although no territorial or state laws of Missouri explicitly authorized slavery, the practice had existed under Spanish and French authority and continued thereafter. In 1857 the Missouri Supreme Court ruled that no legislation was necessary to support the institution.[16] Missouri also passed punitive laws designed to keep free blacks out of the state. By 1860, St. Louis was the only county in Missouri with as many as

one hundred free blacks—at the same time that Illinois had twenty counties
in that category, yet had an African-American population less than one-fif-
teenth that of Missouri.

Although slavery was not totally forbidden in the laws of Illinois and
Indiana, nor was it created by law in Missouri, the discouraging effect which
the Ordinance of 1787 had on the prospects for slavery in the Old Northwest
has to be credited with setting a clear geography of alternatives.[17] Slaveholders who wanted to move west from Kentucky or Tennessee had Missouri as
their option. Others from the same states who wished to avoid slavery—perhaps even the very presence of the African-American—could move north of
the Ohio River. The rapid growth of the Sangamon country and the lower
Illinois Valley was based on the same population source (mainly, the Bluegrass of Kentucky) that fed Missouri. Kentucky's population was about 18%
slave in 1800, and so was Missouri's by 1830. But Sangamon County, Illinois,
counted a total of only 13 slaves in a population of 12,960 in 1830.

Transformation of the old French slave economy into its Anglo-American
successor began immediately in Missouri. Slave populations decreased after
1820 in the French settlements south of St. Louis, where Kentuckians and
Tennesseeans purchased more slaves and headed northwest to the region that
became known as Little Dixie. Missouri's Dixie, a notable outlier of slavery
that is clearly visible on maps of slave numbers beginning in 1820, was separated from other slaveholding areas by the rugged Ozark Highlands.

Good alluvial bottomlands line the lower Missouri River, but away from
the bottoms the topography is mostly rolling. West of the mouth of the
Osage River (present-day Jefferson City) the Ozark escarpment retreats toward the south and prairies appear on the loess-mantled uplands north of
the Missouri.[18] Daniel Boone came to the area in 1807 to make salt. The
surrounding territory, which became known as the Boonslick, started receiving settlers in substantial numbers after the War of 1812 (fig. 28). Slaves were
brought to the Boonslick and to the Mississippi River counties north of St.
Louis. From these two centers settlement and slavery eventually expanded to
form a continuous agricultural region across the north-central section of the
state. Corn-livestock agriculture was established early, and two other agricultural specialties of Kentucky, hemp and tobacco, were transferred to Missouri
as well. Little Dixie was thus a westward projection of the Bluegrass, complete with an emphasis on slave labor and staple crop production. It offered
proof that slavery could exist in a climate that was both cooler and drier than
that of the Plantation South.[19]

A variety of statistics might be used to show that slavery in Little Dixie
was less important and less successful than in its Kentucky parent. Seeking
to demonstrate the point, Jonas Viles stated that in 1860 "only 80" of the 95
Missouri farms of more than one thousand acres were located in Little Dixie,

Fig. 28. Rolling farmland in the Boonslick, Howard County, Missouri. In the 1810s Kentuckians moved to this part of central Missouri that physically resembles the Outer Bluegrass, and they established an agricultural economy somewhat like the one they had left.

and he added, "these figures hardly suggest the prevalence of large scale production."[20] To Viles, slavery in Missouri was a social rather than an economic institution. He noted that in Missouri's slave counties "the bulk of the farm income clearly came from cereals and stock," evidence that slavery was relatively small in scale. While some have argued that slavery was on the decline in Missouri before the Civil War, Perry McCandless noted the continued migration of slaves to the state during the 1850s as well as their increasing monetary value and concluded, "slavery was not a 'dying' institution in Missouri" in 1860.[21]

Missouri's share of the slaves held in the three states of Missouri, Kentucky, and Tennessee grew steadily, reaching a maximum 18.7% of the three-state total in 1860. Blacks accounted for only 10% of Missouri's 1860 population, however, down from a maximum of 18% in 1830. The rate of increase in Missouri's free-colored population was roughly equal to the increase in the state's slave population during the 1850s, despite laws which made manumission unattractive and severely circumscribed the rights of free blacks. Yet slavery was not on the decline in terms of numbers of slaves employed, their value, or the geographical extent of slave territory within the

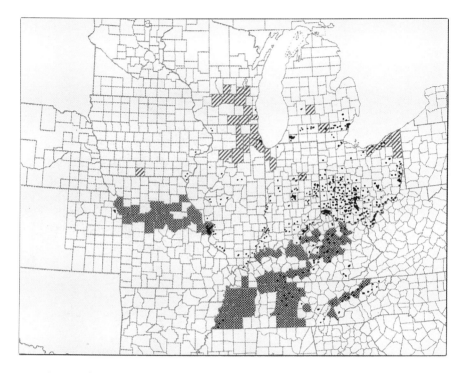

Fig. 29. The ninety-nine largest slaveholding counties (shown in dark shading) in Missouri, Tennessee, and Kentucky in 1860. One dot represents 100 free Negroes the same year. Counties with striped pattern are those in which a third-party presidential candidate received either the largest or second-largest vote in one or both elections, 1848 and 1852. The Corn Belt lay mainly north of slavery, south of the abolitionist strongholds, and, except for parts of Ohio and Indiana, had only a small African-American population.

state: all three of those indicators continued to increase during the late years before emancipation. Nor did the fact that slaves were used in corn/livestock production distinguish Little Dixie from the Bluegrass, the Pennyroyal Plateau, or the Nashville Basin.

The ninety-nine largest slaveholding counties of Tennessee, Kentucky, and Missouri in 1860 (containing at least 950 male slaves or 2,000 slaves in total) were distributed across the major agricultural districts of the three states (fig. 29; table 1). Twenty-one Missouri counties, mostly in Little Dixie, are included partly because the size of their slave populations was a function of their comparatively large areas. Twenty-five Bluegrass counties, twenty-one on the Pennyroyal Plateau, eighteen Nashville Basin or east Tennessee valley counties, and fourteen west Tennessee tobacco/cotton counties complete the list. Correlation coefficients between numbers of male slaves and acres of

Table 1. Slavery and Agriculture, 1860

| Region | Number of counties | Male slaves (x1000) | Statistics per acre improved farmland | | | | Bushels of corn per head hogs and cattle |
			Corn (bushels)	Tobacco (lbs.)	Hemp (lbs.)	Cotton (bales)	
Little Dixie	21	15.1	11.1	6.9	11.8	–	26.4
Bluegrass	25	18.2	7.7	1.9	33.8	–	24.4
Pennyroyal/ Western Ky.	21	22.5	8.8	24.8	–	–	20.5
Nashville Basin/ East Tenn.	18	22.9	7.9	3.5	–	16.3	20.8
West Tenn.	14	50.1	10.8	20.3	–	156.7	15.9

Source: U.S. Census, 1860. Statistics for the 99 largest slaveholding counties of Tennessee, Kentucky, and Missouri.

Table 2. Correlation Coefficients* between Numbers of Male Slaves and:

Region	Improved farm acres	Corn (bushels)	Tobacco (lbs.)	Hemp (lbs.)	Cotton (bales)
Little Dixie	.502*	.515*	.090	.535*	–
Bluegrass	.581*	.639*	– .172	– .045	–
Pennyroyal/Western Ky.	.898*	.772*	.768*	–	–
Nashville Basin/East Tenn.	.733*	.793*	.262	–	.650*
West Tenn.	.995*	.354	– .376	–	.734*

*Significant from zero at .01 level.

improved farmland are strong and positive in all five regions (table 2). West Tennessee, where land clearing and farm-making activities were important in the 1850s, shows a near-perfect correlation between slave numbers and improved acres.

Four of these five slaveholding regions were part of the Corn Belt in 1860; West Tennessee, which produced nearly as much corn per improved acre as Little Dixie, does not qualify according to its ratio of corn production to meat animals (below the 18.5 bushels per animal used as a lower limit for Corn Belt counties in this study). West Tennessee's hogs were more likely mast fed, and its large corn crop may have gone directly for human consumption or it may have entered the Mississippi River trade as cash grain. Corn production correlates significantly with the number of male slaves in the other four regions, and in fact shows a stronger relationship with slave numbers than does any other crop in the Bluegrass, Pennyroyal, and Nashville Basin/East Tennessee clusters. The association is weakest in Missouri, although the correlation coefficient is significant from zero there as well. These relationships suggest that slave geography may even have been less determined by the corn crop in Missouri than was true in the Bluegrass, the Pennyroyal, or the Nashville Basin.

Other crops are correlated with slaves in the various staple specialty regions. Tobacco production and slaves are correlated significantly only in the Pennyroyal Plateau and western Kentucky counties. Tobacco was the Pennyroyal's staple crop, and its slave economy was expanding at the same time that west Tennessee's large slaveholdings were becoming increasingly specialized in cotton. As expected, slave numbers are significantly correlated with cotton production in the two Tennessee regions that grew the crop. Hemp production in 1859 correlates with slave numbers only in Missouri.

Hemp, *Cannabis sativa*, is a rank weed that grows luxuriantly in the hot, damp summers of the southern Middle West. Rope made from hemp was used widely, especially for cotton baling in the South, and its price made production attractive until after the Civil War. The crop was harvested in the fall and allowed to rot in the field into the early winter. Slaves were used in the planting and harvest; even more importantly, they provided the hard labor required to strip usable fibers from the plant. A task system was used in the Missouri hemp counties, where slaves were required to break one hundred pounds of hemp fiber per day. Many slaves apparently exceeded the quota.[22] Although hemp is most associated with slavery in Little Dixie, hemp production per county was greater in the heart of the Bluegrass; and tobacco provided more income than did hemp in Missouri's top slaveholding counties.

Hemp, cotton, and tobacco were the staple crops that made slavery profitable in what otherwise might have been typical Corn Belt counties with free-labor economies. Ralph V. Anderson and Robert E. Gallman have argued that the antebellum Plantation South was able to raise food crops and livestock because those activities utilized labor that was otherwise insufficiently employed in staple production.[23] The converse holds true of these more northern slaveholding counties: the labor required for corn/livestock agriculture was not enough to employ many slaves, but the addition of staples like hemp, tobacco, or cotton consumed enough labor to make slavery an attractive alternative. Corn was produced in all of the staple regions examined here and, except in west Tennessee, which was not part of the Corn Belt, corn production and slave holdings exhibit a significant, positive correlation. Corn was practically ubiquitous, but various staple crops were grown intensively in some counties, yet scarcely at all in others.

Another way to examine the nature of slavery is to compare size distributions of slaveholdings in which farms or plantations are classified in terms of the numbers of slaves per holding. This approach is preferable to comparing the average number of slaves per unit, because the size distribution is typically skewed, causing the average value to be inflated by the relatively small number of large holdings. If each county is treated as though it were a sample of a larger population, then counties can be compared directly, using a test for significant differences between size-distributions across counties (table 3).

For example, Howard County, which includes the early Boonslick settlement in Missouri, and Lafayette County, to its west, had size distributions of slaveholdings in 1860 which were statistically equivalent to those of Fayette and Bourbon counties, Kentucky, in the heart of the Bluegrass region. All four also were statistically similar to Todd County in the tobacco-planting Pennyroyal. In turn, Todd's distribution resembled Montgomery County's, in Tennessee's portion of the Pennyroyal, and those of Maury and Rutherford counties in the Nashville Basin. The two Missouri counties are most similar to the Bluegrass, less so to the Pennyroyal, and different from the two Nashville Basin counties where holdings were larger.

Table 3. Percentage of Slaveholders by Number of Slaves,* Selected Counties, by Region, 1860

Number of slaves	Little Dixie		Bluegrass		Pennyroyal		Nashville Basin	
	Lafayette Mo.	Howard Mo.	Fayette Ky.	Bourbon Ky.	Todd Ky.	Montgomery Tenn.	Maury Tenn.	Rutherford Tenn.
1	13.5	18.7	14.8	15.5	15.0	14.1	16.7	18.2
2–4	26.3	26.7	28.0	26.0	24.2	25.9	26.1	23.5
5–9	33.6	27.5	27.4	30.2	27.0	26.0	24.4	23.9
10–19	15.3	21.5	21.6	22.8	21.3	19.4	20.1	20.4
20–49	10.5	5.2	7.7	5.1	11.6	13.3	11.1	12.4
50–99	0.7	0.4	0.5	0.2	0.8	1.2	1.5	1.5
100–199	0.1	–	–	0.1	–	0.1	0.1	0.2
200–299	–	–	–	–	–	–	–	–

*All adjacent columns are statistically equal, Kolmogorov-Smirnov two-sample test, .05 significance. (Lafayette, Howard, Fayette, Bourbon, Todd) and (Todd, Montgomery, Maury, Rutherford) form groups of statistical equality.

Slavery in Little Dixie thus showed a resemblance to the areas from which it had been introduced to Missouri. Slavery may have been as successful in these Missouri counties as it was simultaneously in Kentucky, and thus it could be argued from this evidence that the institution was exportable to the west. Where cash crops like tobacco or hemp were minor, however, Missouri slaveholdings were comparatively small in scale. Slaves worked on many productive corn/livestock farms in the state, just as they did in Tennessee and Kentucky, but at the county scale, at least, the association between slaves and Corn Belt agriculture is obviously weakened if staple crops are removed from the mix. Slavery seems to have been more in the Corn Belt than of it.

Corn/livestock farming could make use of hired labor but, apart from the work of cultivating corn, the system was not labor-intensive. The corn shock, while perhaps cumbersome to erect, can be seen as a labor-saving device. Fattening animals required comparatively little labor and allowed droving to be an off-season occupation.[24] Compared with seeding, transplanting, hoeing, suckering, harvesting, hanging, and tying down tobacco— or stripping hemp—corn/livestock agriculture was easy work. Regardless of the price differences between slave labor versus free, an agricultural system that seems to have been predicated on labor efficiency is an unlikely candidate to expand in one region rather than another because of differences in labor costs.

Corn Belt agriculture spread north and west across Illinois and Indiana, leaving behind the proslavery sentiments held in the southern portions of those states. Slavery may have been opposed as a degradation of free labor, yet feelings against the presence of free African-Americans continued to deepen across the Corn Belt at mid-century. The populace made its opinion known in several referenda and constitutional conventions held in various states. Eugene Berwanger's data show that, of the various votes taken on exclusion in the states of Illinois, Indiana, and Kansas, 79.1% of the people voted to bar blacks from their states. Iowa's legislature passed a free-negro exclusion bill in 1851, after statehood had been won. Similar bills were introduced in Nebraska in 1859 but failed to pass. At the start of the Civil War, blacks could not vote in any Corn Belt state. Kansas established segregated schools, while in Illinois and Indiana there were no provisions at all for the education of African-American children.[25]

The free-colored population in 1860 was concentrated mainly in towns and cities (fig. 29). More than half of Missouri free blacks lived in St. Louis County, which seems to have become a relatively safe haven for the small number of manumitted slaves in the state. Cincinnati's 4,000 free blacks was the largest concentration in 1860, roughly equal to the combined total of St. Louis, Chicago, and Detroit. Chillicothe's free black population, probably the first in the Old Northwest, started when manumitted slaves were brought to the Virginia District by their former owners after 1800. Ross County, Ohio, counted 126 free Negroes in 1810. The population grew slowly and steadily, mainly through natural increase and local migration, to a total of 2,781 persons in 1860. Chillicothe and its surrounding farms encompassed the second-largest concentration of free Negroes before the Civil War. Ohio had its own code of black laws, but it was the only state in the Old Northwest where blacks accounted for as much as one percent of the population in 1860. Although black populations were increasing in the Northern states, and at a rate large enough to suggest that net migration was augmenting natural increase, the percentage of blacks in Illinois's and Indiana's populations peaked by 1810 and declined thereafter.

Earle's interpretation of the increasing efficiency of slave labor during the 1850s led him to conclude, "slavery was headed north," into Indiana and Illinois at that time. While his reconstruction of labor costs may be correct, he simplified contemporary attitudes toward African-Americans in the Corn Belt. Earle wrote: "the corn-hog region gave increasingly vocal support to proslavery politics, parties, and legislation and helped force through severely restrictive state laws that curtailed the civil rights of free blacks and led them toward servitude if not enslavement."[26] Indeed, it was not enslavement that was sought. People of the Corn Belt did not want slavery in Nebraska and Kansas, because that would mean the presence of blacks. Exclusion was a manifestation of racism, not an endorsement of slavery.

Lawyers were the principal agitators of Negro exclusion at Illinois's 1847 constitutional convention; they were prominent among those who pushed discriminatory measures at Ohio's 1850 convention as well.[27] Farmers who served as delegates seem to have been less interested in the issue, perhaps because black labor was comparatively unknown to them. The only sections where rural black labor was truly available were the Scioto Valley/Virginia District and the Miami Valley. Small numbers of free Negroes lived where slaves once had, along the Ohio, lower Wabash, and Mississippi rivers; some eastern Indiana Quaker settlements also were a haven for free blacks in 1860. But European migration into the Old Northwest and Iowa overwhelmed the small percentage of African-Americans in the population; only Ohio and Michigan registered small gains of blacks as a percentage of total population during the 1850s.

The combined effects of emancipation and war produced a new migration to the North between 1860 and 1870. Blacks approximately doubled their percentage in Ohio, Indiana, and Illinois during the 1860s. Civil War black migrants to Ohio settled in and around the existing enclaves, primarily in the Virginia District and the Miami Valley. By 1870, blacks formed as large a percentage in Indiana and Illinois as they had forty years earlier. And in those states, too, new immigrants took up residence within the existing scatter of black settlements. By 1870, blacks north of the Ohio River lived principally in cities, large or small, especially in the emerging railroad centers. They also moved to Michigan, Wisconsin, Minnesota, Iowa, and Nebraska during the Civil War, but none of those states had black populations that exceeded 1% of the total in 1870. The feared outpouring of emancipated slaves was little more than a trickle, one that had a slight impact on population geography in the Middle West and produced few new concentrations.

The flight from slavery may have been a long one in some cases, but within free territory the tendency was to cluster near the border. Probably the greatest concentration of blacks in the Old Northwest was found on the banks of the Ohio River, facing Kentucky. The flight of Missouri slaves to Kansas similarly stopped for many at Atchison, Leavenworth, and other communities on the west bank of the Missouri. Although Kansas received ex-slaves from a wide area and registered the largest percentage gain in black population of any northern state during the Civil War, the decrease in black population in western Missouri during the 1860s can account for a substantial amount of the gain in Kansas. St. Louis attracted ex-slaves from Little Dixie and saw a three-fold increase in its African-American population during the 1860s, far greater than the increase in Detroit, Cincinnati, or Chicago.

Kentucky's black population declined during the 1860s, Missouri's remained stable, and Tennessee's increased. Black populations declined wherever there were no cities to absorb the local flight from the land. Thousands of ex-slaves left the Bluegrass and typically either moved to Louisville or exited the state. Continued expansion of cotton in West Tennessee produced some rearrangement of blacks within the state, but both Tennessee and Kentucky, like Missouri, found nearly the entire increase in black population in the cities. Comparatively little of the interstate black migration of the 1860s was absorbed by the Corn Belt, but the block of New England-New York counties to its north, the great stronghold of abolition in the Middle West, received even less.

The Corn Belt's response to antislavery politics before the Civil War reflected the regional heritage of its population. Yankeeland—northern Ohio and Illinois, southern Michigan and Wisconsin—turned out substantial votes for abolitionist candidates in 1848 and 1852, but few Corn Belt counties showed much interest in the issue until 1854 (fig. 29). Passage of the Kansas-Nebraska Act meant that the legality of slavery would be decided by popular sovereignty in those two new states that lay in the path of westward Corn Belt expansion. General opinion held that Nebraska would become a free state, while Kansas, positioned to receive the westward migration from Missouri, was practically assured as slave.[28] The door to slavery, once thought closed in the North, was thus reopened. Radicals rallied under the "anti-Nebraska" banner and coalesced within a year to form the Republican party.[29]

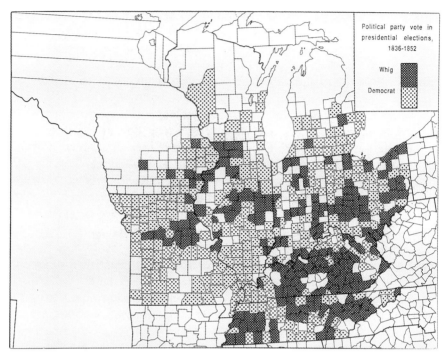

Fig. 30. Political party vote for president in the five presidential elections, 1836–
1852. Shaded counties are those that voted consistently in at least four of the five
elections.

Many of the Corn Belt pioneers of Ohio, Indiana, and Illinois had been
Jacksonian Democrats but voted Whig beginning with that party's formation
in the mid-1830s. In the five presidential elections from 1836 to 1852, won
alternately by a Democrat then a Whig, a rough but generally discernible
regionalization of party vote was apparent in the Corn Belt states (fig. 30).[30]
Little Dixie and the Sangamon country voted Whig, just like that great center of Whiggism, the Bluegrass of Kentucky. The Nashville Basin and the
Pennyroyal also voted Whig, as did most of the Miami Valley and the Virginia District in Ohio. All five of the islands of good land where the Corn
Belt emerged thus aligned primarily with the Whigs.

Poorer lands and frontiers of settlement, whether in southern Michigan
and Wisconsin, northern and western Missouri, or extreme western Kentucky
(Jackson Purchase) voted mostly for the Democrats. The upper Illinois Valley
voted Democrat, the Illinois Military Tract, Whig. Ohio's Western Reserve
was Whiggish, while the belt of Pennsylvania-settled counties across northern
Ohio was Democratic. Economic issues, revolving around the role of national

banking and the level of the tariff, commonly are used to explain this pattern of regional concentrations in party preference.[31]

The Whigs collapsed when the Kansas-Nebraska issue split the northern and southern sections of their party. While many Whigs became Republicans, that was not the only source of support in the cradle of Republicanism, the block of Yankee counties north of the Corn Belt. There, Republicans came from the much larger ranks of Democrats and from third-party abolitionists and Free Soilers. The abolitionist-Yankee stronghold can be seen in the pattern of counties that offered strong third-party challenges in the presidential elections of 1848 and 1852 (fig. 29). Illinois encompassed both sources of anti-Nebraska (Republican) protest: early mass meetings were held at Rockford, a part of the abolitionist north, and in Springfield and Bloomington, where support came from Corn Belt Whigs.[32]

A new alignment emerged under war. The Virginia District and most of the Miami Valley had voted Whig since 1836. These two areas also had supplied a good share of the colonizing population for the Grand Prairie of Illinois and western Indiana by the 1850s. Both parent and offspring entered the Republican party with the 1856 Fremont-Buchanan election; in the Lincoln-Douglas contest of 1860, another dozen Corn Belt counties in the Ohio-to-Illinois stream went Republican. Here, in the heart of the Corn Belt, the issue was not abolition, but rather what many saw as the Kansas-Nebraska betrayal on the part of Senator Stephen Douglas and the Democrats. For these Corn Belt Whigs, keeping slavery out of Kansas and Nebraska was more a matter of excluding the African-American than it was taking a stand against slavery.

Republicanism advanced to the south, but not without opposition. Away from the abolitionist counties, the movement consisted mostly of a shift in party allegiance from Whig to Republican. Using the 139 Corn Belt counties of Ohio, Indiana, and Illinois in 1860 as the basis for comparison, it appears that the balance of party power shifted only slightly in total. Of the 58 strongly Democratic counties in the five elections, 1836–1852, 45 (77.5%) voted with the Democracy in 1856 and 1860. Thirty-seven of the 49 strong Whig counties had gone Republican by 1860, a roughly equal 75.5%. The Republicans gained 13 former Democratic strongholds, but 12 bastions of Whig support shifted to the Democrats in 1856 and 1860.

Although the county totals changed only slightly, the new alignment was far more sectional than the old one had been (fig. 31). Whigs in the Sangamon country and Little Dixie returned to their Southern roots and thereafter voted with the party of Andrew Jackson. The Grand Prairie and northern Illinois/Indiana, lacking that heritage, stayed with the Republicans. Fifteen Indiana counties that supported Lincoln in 1860 voted for the Democrat Mc-

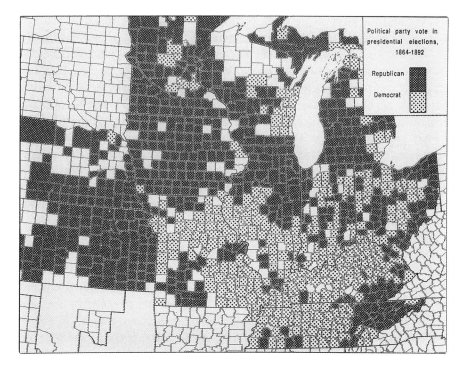

Fig. 31. Party vote for president in eight presidential elections, 1864–1892. Shaded counties are those that voted consistently in at least six of the eight elections.

Clellan four years later, at a time when support for the war was at a low ebb in the state. Eastern Wisconsin, with its large German population, returned to the Democratic column as well.[33] The belt of Pennsylvania-rooted Democratic counties across northern Ohio and Indiana persisted through the war and remained thereafter. Indiana and Ohio retained a pattern of party allegiance similar to that prior to the Civil War, while Illinois and Missouri evolved sectional blocs that closely reflected Kentucky-versus-all-other origins. Northern Illinois, Iowa, Kansas, and Nebraska all went Republican and stayed that way.[34]

The Civil War virtually halted the northward spread of Southern populations into areas not yet settled by whites, guaranteeing that the Missouri-Kansas border would remain as the northwestern corner of the South, the sharpest divide in terms of population origins anywhere in the Corn Belt.[35] Yet even this statement is ambiguous. Although slavery was barely able to penetrate the Kansas border, racism transgressed it easily and found fertile ground. A century after passage of the Kansas-Nebraska Act the United States Supreme Court struck down segregated schools in *Brown* v. *Board of Education*, a suit brought in Topeka, not in the South. Southernness, proslavery and antislavery beliefs, and negrophobia were not one and the same. Each had a geography different from the others, and all overlapped parts of the Corn Belt.

Republican programs of free land in the West and support for railroad construction beyond Council Bluffs became law of the land under the circumstance of secession. Blockade of the lower Mississippi River brought on the demise of the riverboat as the mode for transporting Corn Belt produce to market and helped focus attention on Chicago, the railroad center and rising giant of meat packing. Republicans and railroads helped make possible new legions of yeoman farmers on millions of new acres that came increasingly under the influence of a single city that the War had helped perhaps more than any other. Throughout the Corn Belt, economic life was beginning to take on a new orientation as farmers, tradespeople, bankers, and manufacturers began to focus on a new center in command of the region's economy. This postwar Corn Belt would be Chicago's.

9

Specialization and Westward Expansion

FEWER THAN 150 counties were included in the Corn Belt of 1850, but by 1880 that number had grown to approximately 500 (figs. 3, 4). Expansion of production involved both the spread of corn/livestock agriculture to new lands, whose rapid settlement was encouraged by the Homestead Act of 1862, and the continued intensification of farming in areas where the inclination to feed meat animals on corn was already established. In 1850 more than one-third of the corn had been grown in Kentucky and Tennessee. In 1880 nearly two-thirds of it was produced in Illinois, Iowa, and Missouri, while the two states south of the Ohio River contributed less than one-tenth of the total.

The Corn Belt was moving north and west, its production was increasing, and new specialty zones appeared, based on local factors of accessibility and land resources. Railroad transportation was partly responsible for the Corn Belt's expansion, much more for its intensification. Between 1850 and 1880 Chicago became the overwhelmingly most important packing center, Kansas and Nebraska came into the Corn Belt, cash-corn farming emerged in Illinois, and Texas longhorns entered the cattle trade. By 1880 the Corn Belt was approaching a temporary western limit, challenged by the emerging winter wheat belt of the Great Plains. At century's end the Corn Belt was a large region of increasingly distinct parts, many specializing in hogs, others in cattle, and a few in grain production.

Chicago's rapid rise as metropolis of the Middle West forms one of the more dramatic success stories in American regional development.[1] The city began as a speculative townsite, a modest six-by-ten grid of blocks, laid out in 1830 by the commissioners of the Illinois and Michigan (I&M) Canal Company at the junction of the north and south branches of the Chicago River. Ottawa, LaSalle, and Lockport were similarly created by the canal commissioners between 1830 and 1841 with the intention that those places, too, would attract residents who would build up the territory along the still-un-completed canal's route.[2] Chicago grew from its role as a Great Lakes port, but even that function was limited because the city remained without water access to the Illinois River route down the Mississippi. In 1847 Chicago may have had fewer than twenty thousand inhabitants.[3] Many had been drawn there anticipating the growth that would come once the waterways were connected. For much of its first two decades of existence, Chicago ran on promise.

In 1848 the I&M canal was completed, Chicago's first railroad started operations, Cyrus Hall McCormick began constructing his reaper factory, the city's first livestock market opened for business, and the Chicago Board of Trade was organized. By 1851 the canal was bringing enough corn from downstate Illinois to make Chicago the largest primary corn market in the United States. McCormick's reapers were purchased by farmers, who began shipping more of their wheat to the city, and by 1854 Chicago had become the largest wheat market. Six more railroad lines radiated to the north, east, south, and west of Chicago by 1856. The 1860 census counted more than one hundred thousand inhabitants of the city. In 1862 Chicago passed Cincinnati as the leading hog packer in the United States, and by 1870 it had become the nation's largest cattle market.

Yet throughout Chicago's entire existence as "Hog Butcher for the World," from its rise to prominence in the 1860s until it virtually abandoned the business a century later, swine-raising never really caught on among Chicago-area farmers. The New York/New England-derived people of northern Illinois and southern Wisconsin raised wheat in the early years. After insects, disease, and damp weather caused a succession of failures in the wheat crop, these Yankee farmers often turned to dairying. Unlike Cincinnati, Chicago had no adjacent Miami Valley to supply animals to its packers. Nor did it need one: railroads brought livestock hundreds of miles from the west and south just as the lines aimed east carried stock and manufactured products to market.

The race to build railroads that would connect Chicago with the rest of the nation took the form of a competition between rival groups of entrepreneurs. One result was that railroad terminals were scattered over the city. Because the initial objective was to get to Chicago, rather than through it, only later was there any serious attention given to linking the railroads together in an efficient manner. Livestock was the most important traffic coming to the city by rail and then shipped out via the same means. Chicago's first stockyards were built along Lake Michigan on the south side of the city in 1856, in a location inaccessible to most of the entering railroads. As more lines of track reached Chicago, more stockyards appeared along them, and by 1864 there were a half dozen such facilities in the city, all with the same purpose of receiving Corn Belt livestock either for consignment to Chicago packers or for reloading and shipment to the east.

The Chicago Union Stock Yards opened for business December 25, 1865 (fig. 32). Built on 320 acres of swampy land about five miles south of the center of the city, the enormous facility, unprecedented in scope, served all of Chicago's railroads.[4] Business volume soon repaid the optimism shown by its builders. In 1848, a mere 20,000 hogs and 10,000 cattle had been driven into Chicago. In its first year of operation, Union Stock Yards received

Fig. 32. Chicago Union Stock Yards' stone gate, all that remains of the yards apart from some rails of the stockyards' loop railroad embedded in the pavement.

960,000 hogs, 393,000 cattle, and 207,000 sheep. Approximately two-thirds of the hogs originated in Illinois, and most of the rest were shipped in from Iowa. Cattle came from as far west as Nebraska and Kansas, although Illinois and Iowa cattle feeders contributed more than half the total receipts. Such a massing of livestock in Chicago also helped stimulate larger stockyards in eastern cities such as Buffalo and Albany, where animals relayed via Chicago were sold to packers.

By 1880, eight million head of livestock were received at Chicago. The total increased to thirteen million in 1890.[5] In that year Chicago accounted for one-half of the value added by manufacture in the urban wholesale meat business of the United States.[6] One might imagine that every second pork chop or beefsteak consumed by urban Americans in 1890 once resided, however briefly, in a stock pen on the south side of Chicago. Typical of the product cycle in most industries, meat packing in Chicago attracted a growing number of small entrepreneurs in the early years. The list of eight packers operating in 1851 had grown to fifty-eight in 1864. In the late 1860s, three dozen companies packed hogs and nine handled cattle.[7] The 1890 census recorded forty-one packers in Chicago; Philadelphia reported twice as many packing houses that year, yet it produced only one-twentieth the output of

Chicago. Concentration of the packing industry at Chicago accompanied the rapid evolution of oligopolistic control by just four of those forty-one packers so that by the late 1880s the firms of Hammond, Morris, Armour, and Swift slaughtered, either at Chicago or at branch plants, nearly half of the nation's total meat supply.[8]

Philip D. Armour and Gustavus F. Swift played a major role in organizing the meat industry of late nineteenth-century Chicago and thereby also created the model followed in other Western cities in later years. New York-born Armour began a grain commission business with his brothers at Chicago in 1856, became involved in meat packing with John Plankinton in Milwaukee in 1859, and started packing hogs at Chicago in 1867. Cattle were added to the Armour and Company operation the next year. But no sooner had Armour established himself at Chicago than he turned his attention westward, toward the supply of cattle, and in 1871 he and Plankinton built a packinghouse at Kansas City. The Armour firm steadily expanded its Chicago facilities and also built new plants in Omaha, East St. Louis, St. Joseph, and Fort Worth.[9] Thus, even as Chicago rose to prominence in meat packing, those who guided the industry were pushing their operations toward the western fringe of the Corn Belt.

G. F. Swift was a successful Massachusetts cattle buyer who came to Union Stock Yards, Chicago, in 1875 to purchase cattle. Seeing the possibilities, Swift acquired a Chicago packing plant and eventually followed Armour's lead in branching to the west. Swift first expanded to Kansas City (1888) and between 1890 and 1902 built plants in Omaha, St. Louis, St. Joseph, St. Paul, and Fort Worth. While Armour was the first of Chicago's beef barons to gain nationwide attention for the business skills he showed in reorganizing the meat industry, Swift earned the nickname, "Dressed Beef King," which identified him with a crucial development without which neither he, nor Armour, nor any of the other large packers would have been able to organize the business as they did.[10]

Long-distance marketing of dressed meat became feasible upon the adoption of the railroad refrigerator car, an innovation that Swift, especially, urged on the industry. The river-based phase of meat packing had been predicated almost entirely on cold-weather butchering and curing of meat products for shipment. The industry's shift north to Chicago took advantage of colder winters in more northern latitudes, which extended the winter packing season and, more importantly, made it possible to gain easy access to quantities of ice. Ice houses became natural adjuncts to meat-packing plants in the early 1850s. By 1857 the once-seasonal meat business had become a year-around occupation at Chicago. Experiments with mechanical refrigeration began in the 1860s, and soon refrigeration equipment was being installed in the packing houses. But it was not until the mid-1870s that an effective, economical means was found to refrigerate and ventilate a railroad car loaded with highly perishable, fresh meat.[11]

Chicago's Union Stock Yards might not have been designed the way they were had the possibility of shipping dressed meat been foreseen. Two-thirds of the cattle that came to Chicago's stockyards had to be reloaded for shipment to the East because the butchering of meat intended for fresh sale still was bound to the place where the meat was consumed. Swift became interested in the possibilities of supplying dressed beef, rather than cattle, to the Massachusetts markets he knew so well. He and his brothers, who had remained in the East, undertook the first attempts to win consumer acceptance of refrigerated meat shipped in from a distance.

Swift and other packers in Chicago and Kansas City faced the stiff opposition of Eastern butchers, railroad officials, and stockyards owners, most of whom wished to continue the practice of shipping live animals. The public overcame its prejudice against refrigerated meat, but the business had to be forced upon the railroads. For the entirety of its century-long history as a means of conveying perishable commodities, the typical refrigerated car was owned by a packer or shipper rather than by a railroad company. The major packers acquired their own fleets of meat refrigerators, and they earned additional revenue by renting them to the smaller packers and by charging the railroads a per diem rate for their use.[12]

In the days when cattle and hogs had walked to the shambles, and when rivers offered the only economical means for transporting heavy goods, the packinghouse was located along a river or canal. Railroads freed packinghouses from this locational restriction by their ability to deliver live animals from any origin and dispatch manufactured products to any destination. Refrigerated transport further enhanced locational options by making it possible to ship dressed meat over long distances. Meat packers seized both of these shifts in the logistics of their industry by positioning themselves ever closer to the supply of animals. Big-city union stockyards were only an intermediate step in the evolution of meat packing's geography, because the packer's best interest focused on obtaining animals that were as recently removed from the range or feedlot as possible. What permitted Chicago's rise as a packing center was thus exactly the same as permitted Omaha's or Sioux City's. And as the Corn Belt moved west, so did the packing industry.

Completion of the I&M Canal in 1848 enabled Chicago's packers to ship their products down the Mississippi River system, and this they did, even though the river phase of the industry drew to a close a decade or so following the canal's completion. The canal had only a minor impact on meat packing. But since grain can be shipped by water as easily as by rail it was the I&M Canal that first gave Chicago a central role in the marketing of corn and wheat raised in the Illinois Valley. Access to Chicago meant that corn that had moved downriver could now enter the Great Lakes trade. Farmers increasingly found that they had the option of either feeding grain to livestock or selling it for cash.

When the Chicago Board of Trade was organized in 1848, the city's grain trade was only beginning; board members seem to have occupied themselves passing resolutions concerning the enhancement of commerce in the local area.[13] Even though Chicago became the largest primary market for both corn and wheat in the 1850s, this role was a limited one that mostly involved unloading, storing, and reloading grain for shipment by rail or lake boat to eastern cities. Grain prices fluctuated according to the season. Corn was harvested in the fall; supplies were large at that time and prices were low. Farmers could get a better price for their grain if they withheld it until summer, when elevator stocks were low. Telegraph lines that linked Chicago and New York by 1848 made it easier to arrange contracts between buyer and seller, although instantaneous communication magnified swings in prices. Eastern buyers became increasingly concerned with regularizing this system so that they could expect timely deliveries and not remain at the mercy of unpredictable price fluctuations.

The Board of Trade's grain exchange attempted to overcome the problem by instituting "forward" or "to arrive" contracts that locked in a specified price between individual buyers and sellers. Standardized futures contracts were introduced gradually in the 1860s, prompted in part by the increased trade through Chicago brought about by the Civil War. Futures trading helped stabilize the price of corn, and that led more farmers, especially in the nearby Grand Prairie, to sell at least part of their corn on the Chicago market. By the 1860s approximately 10% of Illinois's corn crop was sold for cash in Chicago.

The futures market is one in which legal agreements are made to buy (take delivery) or sell (make delivery) of a specified amount of some commodity on a given date. Grain growers typically are hedgers in such a market, because they seek to establish a price for a crop, even before it is harvested, by selling futures contracts. Hedgers sell to legions of speculators, the risk-takers who gamble that they will be able to turn over their contracts when the price is right. Hedgers and speculators obviously depend on one another; both are influenced by the world's supply of and demand for the commodity they are trading. It is the transactions between hedgers and speculators that allow prices to be discovered. The invention of futures trading promoted cash-grain farming by giving farmers and grain dealers an insurance policy against adverse price changes.

Cash-grain farming refers to the practice of raising crops deliberately larger than needed for home consumption with the intention of selling the surplus in a grain market. Numerous mid-nineteenth-century references confirm that corn was sold for cash in the Ohio and Mississippi valleys whenever there was a surplus and the price was high enough to bear transportation to a market. In the era of river commerce, farmers who sold corn for cash were those located along or near the rivers. Others were too distant, given the price they could obtain for their corn versus the cost of transportation. The first cash-corn region of the Middle West was a fluctuating one, confined to a narrow zone along any navigable stream. It was overland distance to the river, rather than distance upriver from the market, that limited selling for cash. Railroads removed this restriction. Cash-corn farming emerged as a major subregional variant of Corn Belt agriculture in eastern Illinois during the first three decades following the Civil War, but the circumstances that caused farmers to make such a shift in their operations involved far more than just improved transportation access.

Farmers everywhere, throughout history, have sought crops that can be raised for sale off the farm. Many farmers of the antebellum Corn Belt raised wheat in rotation with corn and oats. They either sold their wheat to a local flouring mill or, when possible, sent it to market, where they hoped it would fetch a good price. Wheat harvests were poor in central Illinois between 1855 and 1860, and this led many to resume their search for a reliable cash crop. One enthusiasm of that period was for raising Chinese sugar cane (sorgo). Upland cotton had long been grown on a small scale in southern Illinois, and some urged its greater adoption, especially during the Civil War; others tried raising hemp, castor beans, broom corn, or sugar beets; some attempted to make Illinois a tobacco-growing state.[14] What all such efforts shared in common was the need to find a new cash crop, especially one that would be more reliable than wheat had been.

Corn was the productive crop for central Illinois farmers, and they needed little urging to plant more of it. The first evidence of a substantial corn surplus emerged in the 1860 census (1859 crop year), when eleven Illinois counties reported harvesting more than sixty bushels of corn per head of cattle and hogs, an amount that would exceed local demand for feed (fig. 33). The surplus was accumulated in the counties fringing the Illinois River north of Beardstown and in several along the Illinois Central and Chicago and Alton railroads across the Grand Prairie. The river counties shipped corn to Chicago via the I&M Canal, the railroads hauled it there directly overland, but nearly all of the corn eventually reached the Lake Michigan docks. About 90% of it was loaded aboard Great Lakes vessels, primarily for shipment to Buffalo.[15]

The flow of corn to Buffalo imitated the established pattern in the wheat trade, but in the case of Chicago corn was the more important export. In the fifteen years ending with 1868, only two (1857 and 1859) saw more wheat than corn moving east from Chicago. The 1868 season was also the first in

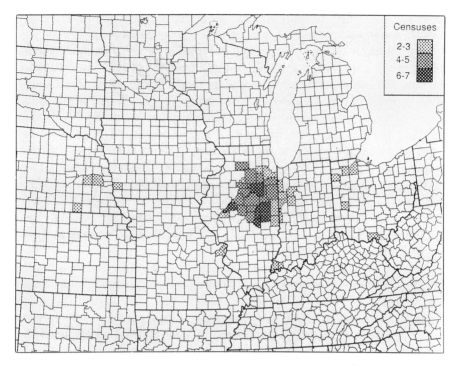

Fig. 33. Cash-grain counties, 1860–1920, defined as the number of census years in which the county produced more than 60 bushels of grain corn per head of hogs and nondairy cattle.

which the I&M Canal was not the largest single source of corn for Chicago's elevators. Inland water transportation thus played a major role in establishing Chicago's cash-corn trade, although in time the far larger available acreage of the landlocked Grand Prairie overwhelmed the possible contribution of the Illinois Valley.[16] But not until 1868 did the Illinois Central Railroad replace the I&M Canal as the major source of cash corn at Chicago.

Annual reports of county correspondents to the Illinois State Agricultural Society during the 1860s and 1870s are filled with the same sorts of discussions about new agricultural methods that appeared in other states. Progressive farmers in Illinois talked hopefully about orchards, flax, sheep, or beekeeping, just like their counterparts elsewhere. Yet it is difficult to find evidence that these same informed observers realized that their region was giving birth to a new system of farming, one that would make all of those promising experiments in diversification insignificant by comparison. Cash-corn farming was adopted gradually by eastern and central Illinois farmers beginning in the late 1850s, not because they learned new methods, but rather because they planted more corn, which was already their main crop, and eventually found they could rely less upon the income received from the sale of cattle and hogs.

The Moultrie County correspondent wrote in 1872, "corn has generally been consumed at home, but is now being shipped to market." An even more reluctant admission of the change appears in that year's Kankakee County entry: "Corn and oats are extensively shipped, though a large amount is fed [which is] the most profitable way of shipping grain to market."[17] By 1876 there was greater awareness, but still little enthusiasm: "The principal resource of Douglas County is the corn crop"; "Ford County is . . . in the northern part of what is known as the corn region." From the Illinois Valley another correspondent declared, "I think Marshall will compare favorably with any county in its shipment of corn," which at least evinces the my-county-is-best attitude farmers normally displayed.[18] These men seem to have been unaware that they were describing an important innovation in Corn Belt agriculture.

Perhaps the change was unremarkable because nothing was new except the allocation of more effort to crops. Feeding corn to livestock on the home farm remained the Corn Belt standard, and thus their own county's increased reliance on the sale of crops may have seemed unimportant, possibly even deviant. Arguments against a decreased emphasis on livestock included the prediction that soil fertility would decrease if livestock manure was not available for use. Isaac Funk's son-in-law, Leonidas Kerrick, an influential McLean County, Illinois, cattle feeder in his own right, thought that "grain farming was but half farming." Speaking before the Ohio State Board of Agriculture, Kerrick exclaimed: "I have pleaded with . . . the people of Illinois on that rich soil, where it is even more difficult to exhaust than here, to come back to their safe and sane methods and combine livestock breeding with farming."[19] Hundreds of eastern Illinois farmers ignored such advice and became specialized corn growers.

The continued increase in the corn crop raised on the Grand Prairie of Illinois and Indiana also can be explained in part by the ongoing efforts at drainage (fig. 34). Draining land requires burying lines of tile, installing surface drains, and digging ditches deep enough to receive the tile conduits that feed into them, arranging the whole system so as to remove excess water from the lowest portion of the fields, yet retaining a water level high enough to allow natural flow into a local stream. Drainage is expensive, it employs substantial labor and machinery, and it requires the services of a skilled surveyor.

Large acreages had been accumulated from railroad grant lands and swamp lands in the 1850s, but the intensity of their use remained moderate. Wet years during the late 1870s had caused widespread crop failure on poorly drained lands, and this focused interest on the drainage problem. The re-

Fig. 34. Drainage ditch on the Grand Prairie in Ford County, Illinois. Despite the destruction of wetland habitats that such ditches wrought, they are interesting environments in their own right and support a variety of plant and animal life.

sponse was swift. By 1882 Illinois had more than one hundred firms engaged in the manufacture and sale of drainage tile; many of them reported themselves unable to meet the demand.[20] Drainage increased the size of the corn crop raised on the wet prairies of Illinois and Indiana, both through the expansion of crop acreage and as a result of the high yields obtained on these seldom-before-cultivated lands.

Tenancy was another stimulus to cash-crop farming in the Grand Prairie. The practice was brought there by men accustomed to such an arrangement, especially in the Virginia Military District of Ohio. In both areas tenancy was looked upon with disfavor by small farmers who believed it was difficult to compete with the large landholders who rented farms to tenants.[21] The money required to drain land was more easily acquired by the large holders—who enjoyed a double benefit because they were able to cover the costs of drainage from the rents their tenants paid and they also could use the ditch-digging labor that tenants provided.

William Scully, born to a family of moderate wealth in Ireland, came to the United States in 1850 and immediately started to purchase lands in Logan County, Illinois. He acquired nearly thirty thousand acres north of Springfield by 1851 and began renting it in small farm units. His first crops were a disappointment. Much of the land was too wet to cultivate, so he tried grazing sheep. Scully commenced drainage experiments in the 1870s, and by 1910 he had improved thousands of acres on his farms. The cost of drainage was met by increasing the rents he collected from his tenants.[22] Scully easily adopted the landlord-tenant mode of the Grand Prairie, and his tenants benefited from the large-scale drainage efforts he was able to undertake. But Scully's tenants, like others, had to increase their own production to pay for the advance in rent. They typically did so by planting more acres in corn, the most valuable crop that could be raised in quantity.

The association between flat, wet prairies, cash-crop production, drainage activities, and tenancy eventually would characterize other areas of the Middle West, although even in 1920 the future was not yet visible in north-central Iowa, which would become a major producer of corn and soybeans in the twentieth century. Eastern Illinois was the only large area that can be said to have specialized in cash-corn production before 1910. Later, place-discriminatory railroad rates on grain would favor this section of Illinois and further reinforce the advantage of selling corn for cash instead of "on the hoof." Wet prairies, when drained, produced large corn crops, but they also demanded a sizable return per acre, because of the increased value of the land brought about through investment in drainage. Tenancy reinforced the need to produce a crop the renter could sell for cash. All of these factors contributed to the emergence of cash-corn agriculture on expensive, drained land.[23]

Iowa was largely unaffected by these developments in the latter half of the nineteenth century. The state's farmers raised corn to feed the hogs and cattle they kept, and only those fortunate enough to have located on river bottoms or on nearby uplands found it remunerative to drive grain wagons down to the riverbank on a regular basis. Hogs were walked to packers in Burlington, Fort Madison, and Keokuk from counties up to one hundred miles inland. To live beyond such a distance did not mean exclusion from the cash economy, however. Farmers in Dallas County, more than one hundred miles from either the Missouri or Mississippi river packers, slaughtered three thousand hogs annually in the mid-1850s and sold them to immigrants arriving in western Iowa.[24] By 1860 several railroads reached across the Mississippi and began drawing Iowa's former southbound river commerce to the east, principally via the trade at stock pens along the track where animals were shipped to Chicago's yards.

The most predictable direction taken by the expanding Corn Belt of the latter nineteenth century was the movement into Nebraska and Kansas. Much of the outrage over the Kansas-Nebraska controversy of 1854 can be understood as concern over what was an obvious expansion path for a free-labor, corn/livestock economy west of the Missouri River. The Platte Purchase of 1837 (the six counties of Missouri north and west of Kansas City) had opened the mid-Missouri Valley to settlement. Large corn crops appeared on the alluvial bottoms, and livestock grazed the rolling loess hills east of the flood-plain; hogs thrived in this prairie-woodland environment. By 1858 the corn-fields had spread north to Sioux City. The Woodbury County correspondent reported, "Home demands for corn have equaled the supply until this fall, and now corn is being purchased by our merchants with view of shipping to St. Louis in the spring."[25] Corn/livestock farmers waited on the east bank of the Missouri River in the 1850s, ready to move to Nebraska or northeastern Kansas.

The Osage Plains of southwestern Missouri also had become part of the Corn Belt by the 1850s. Early settlers gained access to these better lands west of the Ozark Plateau via the Osage River or else came up the White River from Arkansas. The Nashville Basin was an important early source of immigrants. Later migrations to the Osage Plains were rooted in the same sources that supplied southeastern Iowa and the Grand Prairie of Illinois.[26] A preponderance of Kentucky Bluegrass origins, which characterized Little Dixie, therefore did not apply to southwestern Missouri, and neither, apparently, was the latter's population as much attracted to the state because slavery was an option. Only a few thousand slaves were being held in the southwestern counties at the time slavery was widespread in Little Dixie. Many in southwestern Missouri were eager to occupy lands west of the Missouri-Kansas border, which bisects the Osage Plains. Here was a pivotal section of the country that derived significance from being not only the border between east and west, but also between north and south. The Osage Plains became the meeting ground for two almost totally different approaches to the raising and marketing of beef cattle.

The first cattle drives north from Texas were made before the Civil War, although the practice did not become regular until 1866, when some 260,000 head of Texas cattle were moved north across the Red River. The first trail led to the terminus of the Missouri Pacific Railroad at Sedalia, a route infamous for the outlaws who preyed upon any Texas drover daring to enter their domain. "Could the prairies of Southeast Kansas and Southwest Missouri talk," wrote Joseph G. McCoy, "they could tell many a thrilling, blood curdling story of carnage . . . not excelled in the history of any banditta, or the annals of the most bloody savages."[27]

The misdeeds of Missouri outlaws are, in any case, better known than the reactions of Missouri farmers who purchased some of these critters and tried to fatten them for market. The Texas longhorn steer, which lived its entire life on the grassy range of the Southwest, proved a disappointment to farmers accustomed to fattening meat animals on corn (fig. 35). The cattle Missouri farmers had been feeding, like cattle throughout the Corn Belt, had been improved by the introduction of British shorthorn (Durham) stock im-

Fig. 35. Latter-day Texas longhorns grazing the Ft. Niobrara National Wildlife Refuge at Valentine, Nebraska. The longhorn, a range animal poorly suited to corn fattening, posed a brief threat to Corn Belt cattle feeders in the 1870s.

ported to the United States as early as 1783. Improved cattle were taken from Maryland to the Bluegrass and to the Virginia Military District. From there the shorthorn blood line fanned out to the west, penetrating, directly or indirectly, every area where cattle fattening was practiced.[28] Stock from improved herds found their way from Tennessee to northeast Texas, just as they did to the Osage Plains, but improved cattle were a minor fraction of the millions that grazed in all of Texas.

Walter Prescott Webb regarded the free-roaming longhorns found in the Nueces River country as having been essentially feral creatures. They were untended, almost totally ignored, after south Texas was gradually abandoned by Mexican *rancheros* following the defeat at San Jacinto in 1836. In the span of three decades, some tens of thousands of cattle had become several million, yet they still were not being exploited in any systematic fashion. The thinning that resulted from antebellum cattle drives had been more than offset by natural increase. In Webb's rich imagination, the Nueces Valley was "a veritable hive from which the cattle swarmed to the north and west." Estimates of 4.8 million head of cattle on the Texas range have been made for 1860.[29]

Like the razorback hog, the longhorn was a once-domesticated animal brought from Spain that subsequently wandered off to become part of the local fauna. Longhorns were but one variety of the New World land race known as Criollo cattle, derived from the herds that grazed in fifteenth-century Andalusia. Various long-horned Spanish cattle had been taken to the Canary Islands at the time of Columbus's voyages, and from there they went to Hispaniola, where they mingled and multiplied rapidly. Criollo cattle were introduced to Mexico in the early 1520s and subsequently to the rest of Latin America. John Rouse regarded both the Texas longhorn and the Florida "scrub" as breeds that emerged in the nineteenth century: "basically they were Criollo cattle, modified by natural selection during 350 years in North America, on which a small degree of crossing with northern European cattle had been imposed for a few decades."[30] Both breeds were largely unaffected by the splenic fever which was endemic in the subtropical latitudes of Texas and Florida.[31]

But unlike the razorback hog, which was eliminated from the Corn Belt as a result of introducing English swine breeds, the longhorn was inserted into the middle-western cattle business *after* the feeding of improved stock had become an established practice. Corn Belt farmers could laugh at the razorback, but they typically despised the longhorn. Their intruder status is suggested by the terminology of the times: Texas cattle were known as "southern" while the English shorthorn was termed "native." Texas cattle were free to anyone who had the courage to approach them. Bulls and steers were most often driven to market, whereas the cows were culled out and left behind to produce more for the next drive. A select, mature bullock might bring six dollars in Texas but sold for ten times that much in the North.[32]

Thirty-five thousand Texas cattle reached Chicago in 1868, although they were in poor condition after being crowded aboard Mississippi River steamboats from New Orleans to Cairo. Other Texas cattle that came north on riverboats were purchased by some Grand Prairie cattlemen who were determined to feed these exotic creatures to advantage. Arrival of the Texas animals in Illinois was soon followed by contact with local stock; massive cattle mortality resulted.[33] The microscopic tick which carries the fever could not survive in the cold, northern winter, but Texas cattle coming north in warm weather invariably infected middle-western stock that lacked the longhorn's inbred resistance. Although Texas cattlemen disingenuously minimized the danger their tick-carrying cattle brought, Corn Belt farmers correctly identified the source of the fever and sought to ban all Texas cattle from their vicinity. The problem was to recur nearly every year into the 1880s.

Joseph G. McCoy established the first organized market for Texas cattle at Abilene, Kansas, in 1867. He was the youngest of nine children born to David and Mary Kilpatrick McCoy, who had settled on the open prairie of Sangamon County, Illinois, in 1819. The McCoys came from the Georgia-

Carolina Piedmont, the Kilpatricks from the Bluegrass of Kentucky, both typical of the early Sangamon settlers.[34] David McCoy, like many of his neighbors, became a cattle feeder, and his sons Joe, James, and William grew up around the feedlots. John T. Alexander, one of Illinois's largest cattle feeders, was a distant neighbor. In the McCoy's western venture, Joe was the organizer, the one who transformed Abilene into the first "cow town," while James focused his efforts on cattle feeding back home and William handled sales and financial matters in New York.[35]

The fact that the McCoys were a family of ambitious Corn Belt cattle feeders has largely been overlooked in accounts of the Texas cattle era. While any cattle-feeding Corn Belt farmer might have wished to keep the tick-carrying Texas animals away from his herd, it was the specialized feeder who stood to lose the most—or gain the most—depending on how the mass of numbers of Texas beeves was incorporated into the northern meat business. One response was to pass laws that would keep the longhorn at a safe distance; another was to treat Texas as yet another source of stockers for the Corn Belt. John T. Alexander had made (and then lost) a fortune in the Illinois cattle business when the McCoy children were growing up. After the Texas trade commenced, Alexander began large purchases of Texas cattle that he roughed through on grass while he simultaneously fattened shorthorns on corn.[36] He suffered heavy losses as a result of having both types of cattle on his Morgan County farm at the same time.

It was the McCoy's vision to establish a market whereby cattle driven north from Texas or the Cherokee country of Oklahoma could be purchased and then shipped east by rail either to cattle feeders or to slaughterhouses. McCoy's Abilene dominated the Texas trade until 1871, after which time Wichita and Ellsworth, Kansas, became the principal cattle towns. Philip Armour's decision to build a packing plant in Kansas City in 1871 can be attributed to the torrent of Texas cattle pouring into Kansas in those years. By 1873 nearly 2.5 million of them had been driven north to market, although it was clear by then that supply had exceeded demand. Joe McCoy and John Alexander organized the Live Stock Men's National Association at Kansas City in 1873 with the apparent purpose of trying to keep Texas cattle from overwhelming the Corn Belt feedlots. In 1874 McCoy, who had been so instrumental in organizing the trade, wrote, "no more stock cattle are needed or wanted . . . and the sooner the stock men of Texas recognize this fact and cease depleting their stocks at home the better for them."[37]

It is difficult to estimate the number of Texas cattle that found their way into the herds of Kansas farmers, but a substantial assimilation must have taken place in the 1870s. The problem of too many cattle led to new efforts to find an expanded range for their keep. Texas cattle were grazing on the buffalo grass in western Dakota Territory by the mid-1870s, and soon the entire western grassland was drawn into the longhorn's ambit.[38] The succession of new cow towns in later years—Dodge City, Ogallala, Cheyenne, Belle Fourche, Dickinson—traces the expansion west and north (fig. 2).

Corn Belt cattle feeders were less interested in keeping longhorns away from the market than they were in controlling their supply. Prohibitions against entry of any Texas cattle into the Corn Belt did not serve feeders' interests, yet they faced ruin if splenic fever decimated their native herds or if the drives up from Texas swamped the market. A better alternative was to change the nature of the beast. Longhorn cows, it was discovered, could produce quality calves sired by shorthorn bulls. The offspring were good feeders that went to market much heavier than the rangy longhorn, and their costs of production were low.[39]

No solution to the Texas problem could have been more fortunate for the Corn Belt's cattle feeders, whose income came from producing the corn-fattened stock that the market had learned to demand. The threat posed by the sudden availability of an animal unable to efficiently convert corn into animal fat and protein was to be overcome through breeding. Crossing had created the dozens of *Bos taurus* lines found in Europe, from which stock nearly all New World cattle were subsequently derived. Now breeding would be used to eliminate those traits incompatible with fattening on corn, the New World staple.

Great Britain was the source of breeding stock for the western Corn Belt in the late nineteenth century, just as it had been a century earlier when shorthorns were brought to the eastern seaboard. Aberdeen Angus bulls were imported to Kansas in the 1870s, and within a decade many more were purchased by livestock men in the Missouri Valley. Breed records of the American Hereford Association show that between 1848 and 1886 more than thirty-eight hundred Hereford cattle were imported to the United States, the largest share of which arrived during the last six or seven years of that period.[40] Hereford bulls gained popularity in the West after 1880 for the purpose of improving range stock. Hereford breeders in Illinois and Iowa did a brisk business supplying bulls during the 1880s to stockmen from Texas to Wyoming. Hereford (Whiteface) cattle eventually came to typify the Great Plains, while the longhorn, like the bison, was kept largely for nostalgic purposes.

A massive realignment of the cattle business thus resulted from the sudden injection of Texas cattle. Oversupply so depressed the market by the mid-1870s that annual drives to the north were as much in search of pasture as they were aimed at some cow-town along a railroad. Criollo blood was absorbed by intense efforts at improving cattle through breeding. Cattle range, once confined to comparatively small areas like the Virginia Military District or the Grand Prairie, suddenly expanded in area, creating a range from Texas to Montana. Overstocking of the range and a severe winter in 1886 produced a massive die-off of western cattle, but the character of the Great Plains region as a supplier of young stock to the middle-western feedlots had been established.

A unique agricultural specialty region emerged in Kansas during the 1870s. The bluestem prairies of the Flint Hills section of the state coincide to some extent with thin soils developed on a flinty limestone; they are excellent pastures, although less suited for cultivation. River bottoms in the Flint Hills are broad, and they produce large corn crops.[41] It is a unique environment, ideally suited for both pasturing on grass and fattening on corn. Railroad connections to the Southwest made the Flint Hills a convenient stopover for cattle before they were shipped east to feedlots or packers.[42] Of the nearly five hundred counties that comprised the Corn Belt in 1880, only five of them, all in the Flint Hills, kept more cattle than they did hogs.

Kansas's first generation of farmers hailed mostly from the Corn Belt. From the opening of settlement in the 1850s until the arrival of foreign-born groups in the mid-1870s, corn/livestock farming was established in Kansas just as it had been in Iowa two decades earlier. Kansas differed in several ways, having a drier climate and a wider array of population origins. Yankees (who were in the minority, despite their well-known presence in Lawrence during the years of warfare with Missouri) and other wheat raisers broadened the array of agricultural possibilities discussed by early Kansas farmers, but it was feeding corn to swine and cattle that occupied the greatest attention. Early Kansas was a Corn Belt state in the making.

Like other Plains states, Kansas also was settled in part through the efforts of railroad companies that had land for sale. The Atchison, Topeka and Santa Fe (AT&SF) Railway was awarded a three-million-acre grant of alternate sections of land along its line from Emporia to the western edge of Kansas. Land sales began at the company's Topeka headquarters in March 1871, about nine months before the railroad was completed to the Colorado border. The first land purchases were made by Kansas residents, but within six months out-of-state buyers began to arrive in Topeka. Those who entered into contracts with the railroad were mainly residents of the Corn Belt who presumably intended to go to Kansas and practice the type of agriculture they already knew.[43]

One cause of the extended uncertainty as to whether Kansas would be a wheat state or a corn state was the highly variable weather. Few who went there were prepared for the baffling sequence of droughts, deluges, winter thaws, spring freezes, tornadoes, and blizzards they would encounter. Forty-seven million bushels of corn were harvested in Kansas in 1873, the largest crop in the state's then-brief history. In some areas "supply so far exceeded the demand [that] it could only be utilized for fuel."[44] Eastern Kansas farmers were temporarily in the cash-corn business. In 1873 the Leavenworth, Lawrence and Galveston Railroad forwarded 2,650 tons of corn for shipment out of Kansas. Sensing they had a good option to pursue, farmers planted even more corn the following year, but drought and a grasshopper invasion crushed yields to only about one-fourth what they had been the year before. The abundance problem of 1873 turned to real concerns over starvation for some Kansas farmers as the winter of 1874 approached. Many counties called meetings of local leaders to ponder what sort of collective action might be taken to forestall disaster.

Kansas's wheat crop was large enough to feed the state in most years, although in 1873 the same railroad that hauled out the surplus corn brought some five hundred tons of wheat into Kansas. Corn Belt farmers were accustomed to raising winter wheat in rotation with corn, yet they believed the wheat crop was not their principal endeavor. Agricultural leaders in Kansas worried that precipitation was insufficient for the corn crop in many years, because rainfall was likely to be deficient in the summer months when corn needed it most. More moisture accumulated in the winter, when winter wheat lay dormant in the ground, and winter wheat's early-summer harvest made it immune to hot-weather drought. Spring wheat was planted by some Kansas farmers in the early years, but it needed summer moisture just as corn did, and hence the crop never became popular in Kansas. The wild fluctuations in the corn crop led the State Board of Agriculture to advise: "Don't put all your eggs in the corn basket; put most in the wheat basket; it is safer."[45] The two crops were akin to different parts of a tree; wheat was "like the permanent outgrowing of the tree, [corn] like the yielding branches, beaten back before the tempest."

News of both withering droughts and grasshopper plagues was widespread during 1874, and this may have been what caused nearly one-third of the Corn Belt natives who had made down payments on railroad lands to cancel their contracts. Land was selling slowly, but railroad officials had conceived of another plan that simultaneously led to large land purchases and the prospects for increased wheat production. The guarantee of religious freedom for German Mennonites living north of the Black Sea was about to expire in the 1870s. Prussia's consul to the Mennonites of South Russia visited Kansas in 1873 seeking sites for their relocation. AT&SF officials succeeded in luring these highly regarded wheat farmers to their central Kansas lands, even amidst drought in September 1874.[46] The immigrants purchased land at three to five dollars per acre, paid 10% down, and normally took a full twelve years to complete payment. Mennonite transactions in the AT&SF land record books stand in contrast to many of the American entries, which have "canceled" written across them.

Although definite proof is lacking, it was the Germans from South Russia who had to have introduced most of the strains of hard winter wheat that appeared in Kansas by 1880.[47] Varieties from the Black Sea and Asia Minor were superior to the soft winter strains (of English or Mediterranean origin) that Corn Belt farmers had brought to Kansas. Mennonites initially adopted Corn Belt agriculture, but they later shifted to wheat. They were innovators who seem to have had no foreknowledge that the grain they were bringing was better than that already being sown. Like the carrying of Dent corn by Virginians into the domain of the Northern Flints, the introduction of South Russian winter wheat into central Kansas replicated a planting habit of another place and was more likely fortuitous than planned.

Wheat acreage steadily expanded in Kansas during the 1870s, although yields were highly variable. Winter wheat, whether soft or hard, faces the dangers of winterkill and early spring freezes, a problem to which corn is immune, since it is not planted until warm weather arrives. The seeming superiority of wheat in a semiarid climate thus turned out to be a more complicated matter than was initially supposed. Good corn years often were different from good wheat years. The Kansas wheat boom of the late 1870s terminated abruptly in the drought of 1880, and that was followed by a substantial increase in corn acreage. Wheat regained favor in the early 1890s, only to be eclipsed once again by corn a few years later.[48] In addition to the unpredictable weather, Kansas farmers, like all others, had to worry that a good crop year would glut the market and depress the price.

Wheat production was advanced by the arrival of thousands more ethnic Germans during the 1880s, although American-born Corn Belt farmers were more numerous and also had to have been responsible for the shift from wheat as a rotation crop to wheat as a specialty crop in Kansas. Because wheat typically yields less income per acre than does corn, farmers who raise wheat usually devote a good share of their acreage to the crop; that, in turn, precludes the possibility of a large corn crop, which also means fewer hogs. Corn became a secondary crop over most of Kansas by 1910, although its production remained substantial. Swine feeding persisted in the northern and eastern counties, while wheat and livestock dominated the rest of the state.[49]

Similar developments took place in southeastern Nebraska, where corn/livestock agriculture was established before the Civil War. Two of the Miami Valley's swine bloodlines, the Irish Grazier and the Big China, were brought to southeastern Nebraska, where prize animals were being bred and shown by 1873.[50] The railroad boom of the 1870s helped spread cornfields throughout the eastern half of the state. Good lands along the Platte and Republican river valleys formed a corridor of westward expansion that took the Corn Belt west to the 98th meridian by 1879. Determined farmers kept inching corn westward until in 1889 more than five million bushels of corn were harvested west of the 100th meridian in southwestern Nebraska. Al-

Fig. 36. Midsummer cornfield with extreme soil-moisture deficiency after two-year drought. The plants are stunted and the leaves are curled, the latter a drought adaptation that minimizes transpiration from leaf surfaces. Those who planted corn in semiarid environments eventually experienced these conditions.

though Nebraska's corn crop that year was surpassed only by those of Illinois, Iowa, and Kansas, 1889 proved to be the maximum westward reach of the dryland Corn Belt.

It had been common for fertile, new ground to produce large corn crops, so the fact that these two western states could produce three-fourths as much corn as Iowa and Illinois only confirmed the high expectations of farmers who were pushing corn farther west. Droughts in the 1890s changed these beliefs (fig. 36). The problem was not that the climate of western Nebraska and Kansas was too dry to prevent all varieties of corn from being grown there on a regular basis. The crop had, after all, been grown by Native Americans for a thousand years in areas both hotter and drier to the south and west. The problem in the central Plains came, rather, from the need to plant one or another of the derived Corn Belt Dents for use as a feed grain. The failure of Dent corn in dry years was more responsible than any other factor in checking the westward advance of corn/livestock agriculture. Winter wheat became the cash crop of southern Nebraska as it was in most of Kansas, and corn eventually was replaced by the more drought-tolerant grain sorghums.

The Corn Belt shrank eastward until the 1940s as a result of both of these substitutions. Irrigation water drawn up from the High Plains aquifer would reverse the trend once again in the 1960s, but in the meantime corn/livestock agriculture was effectively shut out of the Great Plains. What was lost on the west, however, was gained slowly on the north as the Corn Belt expanded into the Upper Middle West. The northward shift was more limited in area than the expansion to the west had been, but the introduction of two new crops—hybrid corn and soybeans—would make the northern Corn Belt even more productive than the new lands of the West had been in the late nineteenth century.

10

New Crops and Northward Expansion

WHAT COTTON IS to South Carolina, sugar to Louisiana, tobacco to Kentucky, or wheat to Pennsylvania, pork is to Iowa," claimed one Iowa farmer in the late 1850s.[1] Southeastern Iowa's hogs were not as numerous as the Miami Valley's at that time, although the balance was beginning to shift west. There were five hogs in Ohio for every two in Iowa in 1860. In 1880 Iowa passed Illinois as the top hog-producing state and that year outproduced Ohio two-to-one. Iowa is also the first state in which no significant phase of mast-fed hogs can be detected. Unlike states to the east and south that had a woodland fringe of counties producing more hogs and cattle than they had corn to fatten, Iowa's corn crops were large from the start. Farmers had to rely almost entirely on prairie- or bottomland corn to provide feed for their hogs, because of the comparative lack of mature woodlands over all but the northeastern corner of the state (which was not an early area of swine production). If Iowa was a pork state, it was also assuredly a corn state.

Corn Belt agriculture had been moving north, up the valleys tributary to the Mississippi, for several decades prior to the settlement of northern Iowa. Wherever agriculture was begun by migrants from the five Corn Belt islands and their westward extensions, feedlots and cornfields soon dominated the scene. Interior northern Iowa was settled later than the rest of the state, and the majority of its original settlers came from more northern latitudes where corn-livestock agriculture was not the common mode. But by the 1850s the Corn Belt had spread farther north in Iowa than it had anywhere else. For the first time farmers who were accustomed to wheat culture and a fairly diversified system of farming found they had not-too-distant neighbors who were mostly interested in raising corn to fatten hogs. That it took the Corn Belt twice as long to move the comparatively shorter distance across Iowa, south to north, than it did the longer distance, east to west, might be taken as evidence of cold-weather restraints in northward movement, but regional habits probably were more responsible for the lag. Those who were accustomed to doing things otherwise had to be convinced before they adopted a new style of farming.

White Dent corn (a Corn Belt Dent type that included some Flint ancestry) was reported as "giving the best satisfaction" of the varieties raised in Kossuth County, on the Iowa-Minnesota border, in 1858, about twenty years before the county became part of the Corn Belt.[2] Farmers there used

Northern Flint corn only as a breaking crop (sod corn, planted on newly broken ground). Kossuth County, like the rest of the northern two tiers of Iowa counties, derived most of its early American-born population from New York, northern Illinois, and southern Wisconsin; its most important early crops were wheat and oats. Buchanan County, two tiers south of Kossuth, was a Yankee and German stronghold and a center of wheat production, yet in 1858, when Buchanan's agriculture was less than a decade old, a correspondent reported that farmers there raised "the White Dent, the Yellow Dent, the Yankee corn, and the White Flint." "Of these," he added, "the White and Yellow Dent are most cultivated and yield the best."[3] Buchanan and Kossuth counties' farmers obviously possessed the germ plasm of Corn Belt agriculture in 1858, they had already identified the Corn Belt Dent types as best, yet not until the late 1870s, after several bad crop years, did they abandon their reliance on wheat.

Farmers from Pennsylvania and northern Ohio also had raised large quantities of wheat, and their zone of mixture with New Yorkers across middle to northern Iowa predictably saw large wheat crops until the 1870s. The common varieties of Flint and Dent corn seem to have been practically ubiquitous in Iowa by 1857–1858, when the State Agricultural Society polled its county correspondents concerning the kinds of corn being grown in their localities. By this late date the mixing of races of maize had taken place in the East, yet there was a more local influence spreading Dent corn northward in Iowa as well. Some early farmers sold both seed and stock to immigrants who passed their way heading toward the settlement frontier. Because the direction of Iowa's settlement was generally from southeast to northwest, the Dents first planted in southeastern Iowa spread incrementally to the north in the process of frontier advance.

The shift away from wheat in northern Iowa took place during the 1870s. By 1880 Corn Belt agriculture dominated all of the state except for the northernmost tier of counties. But there the line of northward advance stalled, consuming the better part of three decades moving north another fifty miles (fig. 37). To note that farming habits were mostly responsible for this noticeable lag in northward movement is not to claim that climate has no effect on the growth of corn, however, because quite the opposite is true. The genetic makeup of maize is such that the plant eventually adapts to wherever it is able to grow, but generations of selection are required to reincorporate the traits that are successful in a given locality. Dent corn was the best crop to feed livestock, but it did not perform well in northern latitudes unless crossed with one or another Flint type. Northward movement of the Corn Belt meant the adoption of corn as a feed grain, rather than the dairy farmer's use of it as ensilage or "green corn," and wherever large crops of corn for grain were reported, there too were found substantial increases in hog production.

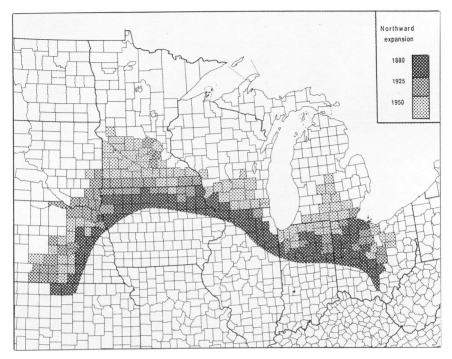

Fig. 37. The Corn Belt spread northward slowly after 1880. (Inclusion in the Corn Belt in 1925 and 1950 is based on production of at least 10 bushels of corn per acre of total cropland, whereas in 1880 the criteria is 7.5 bushels per acre of improved land.)

Oscar H. Will of Bismarck, North Dakota, collected the seeds of Flint and flour corns that had been grown for centuries by the Northern Plains Indians. Will and his son George later tried introducing semi-Dent corn into North Dakota, but the results were poor. George Will wrote, "any effort to acclimate [Dent corn] beyond a certain latitude or perhaps isothermal line results in a rapid shrinking in size and in crop production."[4] He thought that the Flints were "pre-eminently best adapted to cold conditions both of soil and climate." Flints were grown almost exclusively in New England in 1899, and the reported corn yields in Vermont, New Hampshire, and most of Massachusetts were equal to those of the Grand Prairie of Illinois, even though corn-for-grain was scarcely a major crop for New England farmers.[5] Climate did not keep corn—all corn—out of northern latitudes, a fact established by the Indians at the time Europeans settled in the North American mid-continent, but the climates of specific localities did reward some varieties more than others.[6]

Taking Flints north to longer day-lengths makes them grow taller, as does an increase in atmospheric humidity, which scatters more sunlight.[7]

Taking them south has the effect of going back in time. One of the Northern Flint types often planted as a sod-breaking crop in Iowa and Kansas was King Philip, a small, early, eight-rowed yellow. It was grown in Kansas as early as 1872 and was kept pure growing at the State Experiment Station in Manhattan beginning in 1876. By 1888 the station reported that their King Philip was "no longer a Flint corn, while in size and habit of growth it more nearly resembles a medium Dent sort than the familiar New England variety from which it descended."[8] The New England standard, King Philip, like all varieties of maize, had emerged in the slow northward spread from tropical latitudes. The climate of Kansas subsequently selected traits that had not been viable in New England but which, given King Philip's genetic makeup, were ready to be called to the fore.

In the 1850s Virginia Gourdseed corn, a Southern Dent type, was reported to grow well south of latitude 42° (say, Chicago), while the Northern Flints seldom received favorable comment south of latitude 39° (south of Cincinnati).[9] Many factors influence the length of growing season, but latitude totally determines day-length, and this factor was understood to be an important consideration in selecting corn varieties. Illinois farmers were advised they "should seldom go more than 50 miles north or south" when choosing seed; fifty years of cultivation was thought needed to develop a race of corn in a locality.[10] In the late 1850s a store of Kentucky corn was brought north to Iowa, where it was sold for seed and planted the next spring. It produced "a beautiful sound grain, and grew well but did not ripen."[11] Latitudinal limits on corn varieties have even been suggested as a factor influencing the latitudinal pattern of westward migration flows in the Middle West.[12]

Until the commercial development of hybrid corn took place in the 1930s, crosses between the various open-pollinated varieties were practically the only source of increased yields. The band between latitudes 39° and 42°, which included most of the early Corn Belt, can be considered as the approximate zone where Flint-Dent crosses are most likely to have been made in the nineteenth century, although certainly accidental and possibly deliberate crosses had occurred in Virginia (at latitude 37°) by 1705 as Beverley described. Crossing was used to improve corn in Pennsylvania in the 1820s. Whether the U.S. Commissioner of Patents' nationwide dissemination of seeds in the late 1840s extended the range of any type or had more the effect of confirming farmers' already-held beliefs in the superiority of certain varieties, the knowledge that crosses were easy to make became widespread as a result. Gourdseed needed a longer season to mature, and it was planted early; the Flints were planted later. This provided a likely circumstance for accidental crossing, because farmers filled in the hills where Gourdseed did not germinate with the seeds of one or another Flint type.

Southern Dent corn was soft, which made it a poor cash corn because it did not store well, yet it was easily chewed and afforded ready-made feed for either hogs or cattle. New England and New York farmers, in contrast, had devised various means of cracking, grinding, and cooking corn for feed, because their hard, Northern Flints required it. While Virginia Gourdseed was an excellent high-yielding feed grain, it was not a success in the long day-lengths and short growing season of the upper Middle West without any introgression of Northern Flint influence. The more northern latitudes thus tested the success of breeding experiments that sought to extend Dent corn northward without sacrificing its desirable qualities as livestock feed. One of the first northern semi-Dent types, Minnesota 13, was developed at the Minnesota Experiment Station at St. Paul and distributed in 1897. Shortly thereafter Alta, a Yellow Dent, was developed near Highmore, South Dakota.[13] With these developments profitable grain corn production became possible north of latitude 44°, although as in the case of northern Iowa a few decades earlier, mere availability of the crop did not mean that farmers adopted it immediately as a specialty.[14]

The Corn Belt moved north in Indiana, Michigan, Wisconsin, and Minnesota between 1890 and 1950, but the advance was slow compared with the rapid westward expansion of earlier decades. Only in Minnesota, where corn production doubled during the 1890s and doubled again by 1920, was there a significant testing of climatic limits. One stimulus for the shift came from the northward expansion of the meat-packing industry, especially pork packers, to cities of the northern Corn Belt. St. Paul was peripheral to the supply of hogs when the first packing plants were built there in the 1880s, but by 1920 Minnesota produced half as many hogs as Illinois, and South St. Paul's stockyards were the sixth largest in the nation.[15]

Although this early twentieth-century northward expansion of the Corn Belt was the result of many farmers' decisions to shift their production away from mixed farming to emphasize livestock feeding, their ability to make such a change was enhanced by maize genetics. Flint introgression favored a northward movement of the Corn Belt, because this was the direction, genetically as well as geographically, in which the greatest improvement in yields would be obtained through Flint-Dent crosses. But there was no symmetry in the exchange to encourage the spread of improved crosses toward lower latitudes. Beginning with the plant's earliest emergence in tropical latitudes, maize diffusion and the frequent emergence of new varieties had always accompanied the spread toward higher latitudes in both the northern and southern hemispheres. Farmers in the southern Corn Belt, where the Northern Flints seldom performed well, had no comparable prospects for increased production to match those on the northern fringes. The best Kansas farmers could do was to cross a derived Corn Belt Dent with Virginia

Gourdseed or Shoe-peg; "none of the flint varieties" was thought to be "of special benefit to the farmers of Kansas."[16]

By 1900 a clear demarcation in corn yields appeared roughly along latitude 39°. South of that line no county reported yields in excess of forty bushels per acre, and only the better corn districts, such as the lower Wabash Valley, had yields above thirty bushels.[17] New lands produced large corn crops nearly everywhere in the early years of trans-Appalachian settlement, but the initial advantage of regions like the Nashville Basin and the Bluegrass had long since disappeared. Yields stagnated and in some areas declined due to the continuous cropping of corn. Nathaniel Shaler deplored the practices of Bluegrass farmers, but he thought that weathering of the underlying calcareous limestones had released enough nutrients to prevent disaster. "No other land of the world is so fitted to withstand the evils of the utterly unscientific agriculture to which it has been submitted," Shaler wrote in 1888.[18]

The Bluegrass and the valleys of East Tennessee virtually dropped out of the Corn Belt by 1880 (fig. 38). The Pennyroyal began specializing even more heavily in its great staple, tobacco. The Nashville Basin remained of those early centers of the Corn Belt south of the Ohio River, although corn yields lagged behind what they once had been; the decimation of livestock herds there during the Civil War had led many Nashville Basin farmers to shift into cotton.[19] The Missouri Bootheel and the limestone valleys of southern Tennessee raised corn, but it was secondary in importance to the cotton crop in those areas.

While the Corn Belt had been retreating from the droughts of the West in the several decades preceding 1920, the counties that dropped out before 1890 were mainly found in the southern half of the region. In many cases they had been among the earliest counties to adopt intensive corn/livestock agriculture. Worn-out land or poor land-use practices in general could be cited as reasons for their declining production, but the lack of new, higher-yielding corn varieties hit them especially hard. If yields were double in central Iowa what they were in the lower Ohio Valley, it was difficult for the more southern farmers to compete in the same business. A shift to a more mixed agriculture in the southern Corn Belt, including the reincorporation of wheat culture and livestock grazing (rather than feeding), was in part a response to the success that corn production was enjoying in the more northern latitudes.

Variations in soil quality also made a difference. The claypan soil area that extends from the southern Illinois gray lands across northeast Missouri to the central portion of southern Iowa had become part of the Corn Belt by 1860. Soils there have a slowly permeable clay layer beneath the surface that can make the plow zone either too wet or too dry for healthy plant

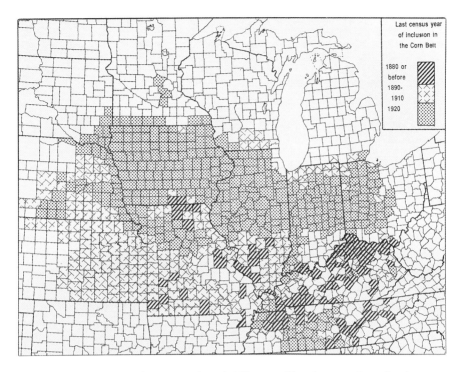

Fig. 38. The Corn Belt's net northward shift was evident by 1920. Counties that had not been included since 1880 were primarily in the southern half of the region. Those that had been included as recently as 1890, 1900, or 1910 were largely on the western fringe, where corn-livestock agriculture was retreating.

growth if rainfall is not well distributed over the season.[20] Counties in the claypan area were at a disadvantage to those farther north that did not have the problem. By 1910 southern Iowa and northeast Missouri farmers were planting less corn and using more of their acreage for pasture. They became suppliers of young cattle and feeder pigs, which were sold to the north, where corn crops were larger and more reliable.

Even in the Grand Prairie and the Wabash Valley farmers worried that new corn varieties weren't boosting yields to the extent that previous introductions had. In the 1920s Corn Belt corn production leveled off, in part a response to price decreases and oversupply, but also because yields were decreasing on lands that had supported large corn crops for many decades. As the limits of the corn-hog production cycle became clearer, farmers sought new options to improve their situation. The enormous corn crops of mid-America had provided cheap feed for hogs, so cheap that the emphasis in swine feeding had generally followed a trend toward heavier animals. Lean, bacon-type hogs were favored between 1890 and 1910. But with that exception, the history of Corn Belt hog production after the 1840s had emphasized heavy, lard-type hogs that consumed as much corn as possible.[21]

Consumer tastes began to shift in the United States after World War I. As in Europe a few decades earlier, the public started to demand leaner cuts of meat. Salt fat pork, mess pork, and heavy hams and shoulders that once dominated the market were no longer desired. American cattlemen soon began marketing baby beef, and the marketing of mutton was reduced in favor of lamb. A 150-pound hog produced more nutrition for the feed than a 450-pound lard hog and yielded lean bacon that brought a higher price.[22] Even more significant for the Corn Belt was the increasing demand for vegetable oils rather than lard.

Corn surpluses became a national problem in the 1920s even as yields were decreasing in some parts of the Corn Belt. Surpluses had appeared occasionally in earlier years but were chronic after the primary depression that followed World War I. Wartime demand pushed the cash price of corn to $1.52 a bushel in 1918. In 1920 farmers harvested more corn than they had during the war, and the price slipped to $.60 a bushel.[23] The price of slaughter hogs declined as well, and for the same reason of oversupply. By the mid-1920s the Corn Belt was acknowledged to have a problem. In an early analysis, Alonzo E. Taylor of the Stanford Food Research Institute wrote, "Hog raising was developed in the corn belt as a part of diversified farming, on the basis of the once correct postulate that raising hogs was the cheapest way of converting grain into fat and the corn belt the most efficient region in which to do this." Taylor even questioned the economic underpinnings of the corn-hog ratio, the device promoted by Henry A. Wallace, which supposedly allowed farmers to predict whether their corn would be more valuable fed to hogs or marketed for cash.[24]

The problem of oversupply and depressed prices only got worse in the 1930s. Under the auspices of the Agricultural Adjustment Administration (AAA), created in 1933 as part of the first New Deal, farmers were eligible for a loan of forty-five cents a bushel against grain stored on their farms if they agreed to reduce corn and hog production. In August and September 1933 the government purchased and killed 6,188,717 pigs and 229,149 piggy sows, which were converted into lard, fertilizer, grease, and salt pork for relief agencies.[25] The 1934 hog crop was about 40% smaller than in previous years, but within five more years hog production returned to former levels, and the government's effort to reduce the pork supply had come to naught.

The time was right to introduce a new cash crop that would break the long-term dependence on corn-livestock production. The crop was soybeans, *Glycine max* (L.) Moench, and they were to provide a triple benefit. The leguminous soybean has a bacterium that attaches to its roots but not to those of any other plant. The bacteria fix atmospheric nitrogen in the soil,

Fig. 39. Soybean fields, July 4, 1991, east of Farnhamville, Iowa, on the Des Moines lobe.

where it becomes available as a nutrient taken up by plants. Corn is a heavy nitrogen feeder, and thus a rotation of corn and soybeans increases the next year's corn yield on the same ground. Soybean roughage provides a usable hay for livestock, and its seeds yield oil that has a high protein content. As early as 1922 agronomists predicted that soybeans would replace oats as a rotation crop in the Corn Belt.[26] This they did, but at a pace even the most enthusiastic promoters did not predict in the 1920s (fig. 39).

Benjamin Franklin was apparently the first to introduce soybeans into the United States; he found the crop in France, where it had been brought from East Asia. Little interest was evident until 1854, when the U.S. Commissioner of Patents had the Perry expedition bring several soybean cultivars from Japan. The seeds were mailed to inquiring farmers, just as the corn varieties had been. Further interest led the U.S. Department of Agriculture to import a number of European and Japanese varieties in 1898.[27] The yellow-seeded soybean, which has a high oil content, was being grown regularly at the North Carolina Agricultural Experiment Station at that time, and the crop had been tried as a means of improving fertility on worn-out cotton and tobacco lands of the Coastal Plain. About seven hundred thousand acres of soybeans were grown in North Carolina, Virginia, and Alabama in 1920, the three states accounting for about three-fourths of the national total. The first mill to crush U.S.-produced soybeans operated at Elizabeth City, North Carolina, in 1912.[28] Soybean crushing could be incorporated into the Southern cotton oil industry, and the crop's restorative effects and forage value were welcomed by a few cotton planters willing to experiment, but it took many years for soybeans to develop from these modest beginnings into a major Coastal Plain crop.

Until World War II, the largest producer of soybeans was China and the leading exporter was Manchuria. Soybeans were successful there in a climate that resembles the central Corn Belt, and this fact was stressed in the early attempts to interest potential soybean growers. While planting corn to feed cattle and hogs was a respectable activity in the Middle West and needed no justification, the idea of planting the leading crop of Manchuria required salesmanship. The first annual meeting of the American Soybean Association took place in a camp-meeting atmosphere September 3, 1920, at "Soyland," the farm of Taylor Fouts and his brothers, Finis and Noah, of Carroll County in the Wabash Valley of Indiana. Purdue University agronomists gave several speeches to the crowd, following which a luncheon featuring baked soybean salad and roasted, salted soybeans was served by the Presbyterian Ladies Aid Society. After lunch "a quartet of local soybean growers sang a very appropriately worded song, 'Growing Soybeans to Get Along.'" The day closed with tours of the Fouts brothers' bean fields, some of which had been planted for forage, others for seed.[29]

Many farmers who tried a few acres of soybeans as a forage crop understood they had another purpose, although the list of "things" made from soybeans could not have seemed very familiar to them: tofu, miso, soy sauce, soap, paint, vegetable cheese, soybean milk, and soy flour for diabetics and babies.[30] An early agronomic text on soybeans included a section of recipes. Soybeans had a cult status even into the 1930s, when their virtues were trumpeted by Henry Ford, who exhorted some of his research chemists to find a way to "grow" an automobile; the Ford Farms in Michigan raised 12,000 acres of soybeans in 1936.[31] Soybeans were championed early by vegetarians who made use of their high protein content in fashioning meat substitutes. Soybeans were a "miracle crop" that produced "gold from the soil," yet the puzzling array of seemingly unrelated end-products derived from them was not reassuring to men whose thoughts ranged from pork to beef.

J. C. Hackleman, an Extension agronomist at the University of Illinois, spread the gospel of soybeans among the state's farmers. In 1922 Hackleman secured the promise of the A. E. Staley Company, a cornstarch manufacturer in Decatur, to crush and process that year's soybean crop growing in central Illinois. Farmers were delighted at the high price that seemed guaranteed that season but were dismayed when the harvest came in and the price dropped. Staley refused to operate the mill without a substantial quantity of soybeans, while farmers held back their supply hoping the price would increase.[32] It took several years before the system operated like other grower-processor markets. Trading in soybean futures began at the Chicago Board of Trade in 1936.

Augustus Eugene Staley, born in the Piedmont of North Carolina in 1867, owned a large cornstarch business in Decatur by the early 1920s. He promoted soybeans with the same enthusiasm he showed for corn products. Although he did not come from the soybean fields of North Carolina, as company publicists claimed, Gene Staley clearly envisioned how important the crop would become at a time when few shared his optimism.[33] The American market for traditional East Asian soy foods was weak, but Staley saw that the demand for soybeans would not be based on those uses. The first big increase in U.S. acreage came when soybeans were successfully introduced into the manufacture of margarine and shortening, as they had been in France and Germany. Corn Belt farmers were better able to grasp the significance of soybean oil meal, the principal by-product of oil extraction, which came into great demand among livestock, poultry, and pet food manufacturers because of its high protein content. In 1940 the Ralston-Purina Company of St. Louis advertised itself as the world's largest consumer of soybean meal.[34]

Dale W. McMillen did come from a soybean-growing area (Van Wert County, Ohio), and he made use of their high protein in the Wayne Feeds that he began manufacturing at Fort Wayne, Indiana, in the 1920s. McMillen's business merged with another feed manufacturer in Peoria in 1929 (the company became Allied Mills), but McMillen returned to northeastern Indiana, where he purchased a defunct beet sugar refinery at Decatur. McMillen formed Central Soya Company (MasterMix Feeds) around a new soybean manufacturing facility he built at Decatur, Indiana, in 1934 and soon constructed another facility at Gibson City, Illinois. Central Soya's Decatur, Indiana, plant was the first in the United States to use the European solvent-extraction method, a superior procedure for oil extraction that replaced the hydraulic and expeller methods used prior to that time.[35] The Archer-Daniels-Midland firm began processing soybeans at Toledo in 1929 and within six years had constructed similar facilities in Chicago, Minneapolis, Milwaukee, and Buffalo. During the 1930s Staley branched to Painesville, Ohio; Ralston-Purina built feed-manufacturing plants in Ohio, Arkansas, and Indiana; and Allied Mills branched to seven more cities, ranging from Portsmouth, Virginia, to Omaha. Two Minneapolis companies, General Mills and Cargill, also entered the soybean business and built plants in Iowa, Minnesota, and Illinois.[36]

Soybeans fit easily into the annual routine of a Corn Belt farm. Their dates of planting and harvest were about the same as for corn. Investment in new equipment could be minor because of attachments that adapted existing farm implements for use with soybeans. But unlike the corn crop, which was then, as now, primarily consumed for livestock feed on the farms where it is grown, soybeans were of no use to the farmer apart from their forage value. For the farmer, beans had to be a cash crop, whereas corn might be. Because soybeans nearly always brought more per bushel—and sometimes more per acre—than corn, they were planted on expensive land. These several factors tended to concentrate soybean production on drained lands within the Corn Belt.

After the mid-1930s, soybean oil and meal accounted for most of the disposition of the crop. The manufacture of shortening and margarine was drawn east toward major cities; meal was sold primarily to manufacturers of animal feed, who were concentrated in the Middle West. Soybean milling began to shift toward the east in the 1930s, partly because of a peculiar structure of tariffs that governed the transportation of unmanufactured agricultural products. Soybeans belong to the class of shipments that have transit privileges, meaning that a quantity shipped from its point of production can be milled along the way and the milled product forwarded to market under the same rate as that charged the raw commodity moving from its point of production to the market.

Transit privileges were promoted by grain millers who wished to protect their investment in cities located somewhere between the grain-growing region and the market (because such intermediate locations otherwise would incur higher transportation costs). Railroads usually lost money when transit privileges were allowed, but they used them to guarantee that both the grain and the milled product moved over their lines. Soybean meal was about as valuable as soybeans, and thus its manufacture enjoyed the transit privilege, while soybean oil was deemed a separate product (a far more valuable one), and it typically had to bear higher freight rates.[37] The net result was to pull the soybean milling industry east toward the market for oil, but this also stimulated interest in growing soybeans wherever a plant was built.

A map of the soybean industry at the end of the 1940s reflects these several factors that had influenced its brief history (fig. 40). Production was concentrated on the best land, especially flat, drained land that could be worked profitably with mechanized equipment. The Maumee Plain of northwestern Ohio and the Des Moines lobe in Iowa both became important soybean producers, as did the alluvial wetlands of the Missouri Bootheel and the St. Francis Basin of Arkansas.[38] Soybean processing plants were located along railroads that could deliver a dependable supply of beans, while each new mill enticed more local farmers to grow the crop. Several railroad companies participated in the soybean boom, especially the New York, Chicago and St. Louis Railroad (Nickel Plate Route), whose line from St. Louis to

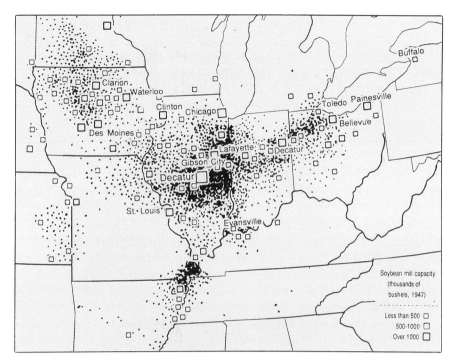

Fig. 40. The soybean industry in the late 1940s. Each dot represents 60,000 bushels of soybeans harvested in 1949 (see chap. 10, note 36).

Toledo and Buffalo became a virtual corridor through the bean fields by the 1940s. No fewer than ten processing plants along its tracks across Indiana and Ohio enjoyed the transit privilege. The map of soybean production formed a giant "Y" across the Middle West. One arm extended northwest and was defined by the availability of drained prairie land in Illinois, Iowa, and Minnesota; the northeast arm grew along the alignment of railroad lines that served soybean processors in Ohio and Indiana; the lower leg was the alluvial Mississippi Valley.

The geography of soybeans reflects a peculiar combination of natural and institutional factors that set this crop apart from the long, regional-historical process that had created the corn/livestock region of the Middle West. Soybeans had no folk associations in the United States. No group of people had traditionally raised the crop, indeed, had even heard of them before 1920, yet by the 1940s soybeans had become established as the second cash crop in the best lands of the Corn Belt. The disappearance of work horses led to a decline in demand for oats, and this freed more land for soybeans. Government programs to restrict corn production freed thousands more acres, and those, too, were often planted in soybeans. Because soybeans would grow wherever corn grew, northward expansion of the Corn Belt was simultaneously an expansion of soybean production.[39] It took Illinois, the first major soybean producer, eleven years to increase its output from one million bushels to ten million; Iowa did it in nine years, Minnesota in six, Ohio in four.[40]

Soybeans were the wonder of the Corn Belt, yet even as their production took order-of-magnitude leaps during the 1930s and 1940s, new developments in corn breeding were drawing ever greater attention. Given the intense interest in producing improved varieties of corn, it is perhaps surprising that hybrid corn did not appear for more than half a century after Gregor Mendel, the Austrian cleric, developed the principles of segregation and independent assortment in genetic inheritance that provided the underlying theory. But Mendel's work was forgotten for several decades. When his results were independently rediscovered later and, ultimately, applied to corn-breeding experiments in the United States, it took years of inbreeding to return to the point where hybrid crosses could be made advantageously. The lag in developing hybrid corn was due more to inertia than to any lack of technical expertise on the part of plant breeders once the principles of Mendelian genetics were known.

Frustrated with all of the new Corn Belt Dent introductions that didn't yield much better than those they replaced, farmers in the early twentieth century had turned their attention to corn aesthetics. Numerous "corn shows" were held around the Middle West in which entries were judged according to a showcard checklist of traits similar to those employed in stock judging. Beautiful ears of corn, when shelled and planted the next year, didn't yield any better than the less attractive, however, and the corn shows eventually fell into disfavor. As a young man Henry A. Wallace had challenged the "pretty ear" corn shows, and later, in the pages of *Wallace's Farmer*, he scorned those who believed in them.[41] But Wallace had no better idea of how to improve corn yields at that time than did most of his contemporaries who also experimented with varietal crosses.

The trend in corn yields had reached a plateau in Illinois by 1900, and any gains after that time were usually of short duration.[42] A group effort to overcome the yield problem was begun in Woodford County in January, 1919. Over one hundred farmers each brought one hundred ears of their best corn to a meeting. The poorest twenty of each hundred were discarded, and a random sample of ten of the remaining eighty was chosen for planting that year. The same procedure was repeated in 1920 and again in 1921. At the end of the 1921 season the highest yield, 68.5 bushels per acre, had been obtained by one of the farmers, George Krug. This variety, which quickly became known as Krug corn, was produced in large quantities for seed by Woodford County growers. By 1928 Krug corn was planted in nearly half the fields of central Illinois; it was the most-planted variety in the state of Iowa.[43] George

Krug's yields hadn't been that much larger than any of his neighbors in the 1919–1921 experiments, but the hope of being able to obtain yields as good or better than those in the trials led thousands of Corn Belt farmers to adopt, within a few years, a single seed corn variety. Krug was the most widely planted open-pollinated corn grown from Ohio to Nebraska for ten years just prior to the general introduction of hybrid corn.

The farmers of Woodford County were following procedures, many of them recommended by experiment station agronomists and described in corn-growing textbooks, that practically guaranteed diminishing returns for their efforts. The farmers did not understand that the reason those early Flint-Dent crosses had produced so well was that the two races of corn were truly different, not just in outward appearance (phenotype) but also in genetic makeup (genotype). The first time that pollen from Flint tassels lodged upon the silk of a Dent ear, hybrid vigor was captured to a maximum degree. But the more crosses that were made among Corn Belt Dents, the less likely it was that the desirable traits would be fixed in succeeding generations. The central Corn Belt was awash in varietal crosses that had been made for years, accidentally or deliberately, between the open-pollinated varieties, and any further introgression of one race into another made it more difficult to isolate the desired traits.[44]

The rediscovery of Mendelian genetics reversed the entire procedure. Edwin Murray East had begun inbreeding experiments at the Illinois experiment station, but he was discouraged by his superiors and left for the Connecticut station in 1905. In 1908 East learned of the work of George Harrison Shull at the Carnegie Institution's research station at Cold Spring Harbor, Long Island, and the two men began debating the role that inbreeding could play in producing better hybrids. It was Shull who saw that inbreeding (or "selfing") would reduce the genetic complexity present in every cornfield. Every plant was a complex hybrid of traits that differed only by chance from the plant next to it.[45] The only way to isolate what was true to one race from the variability it had acquired by chance through years of open pollination was to guarantee that the silk of the plant was fertilized by its own tassel's pollen alone. The inevitable result, within a few generations, was decreased size and an absence of vigor, but each year's seed represented a step in producing a homogeneous genotype. Five to seven generations of selfing was sufficient to eliminate the heterogeneity that years of open pollination had produced.

Thus even as Corn Belt farmers were ordering their open-pollinated Krug corn in the 1920s, hybrid seed corn was being developed that would make it unnecessary to undertake any more programs like the one in Woodford County. One of the Woodford County farmers who grew Krug corn, Lester Pfister of El Paso, stole a march on his neighbors by selfing some of

Fig. 41. Experimental corn plots on the agriculture campus of the University of
Illinois in Urbana.

his plants. Although Pfister also sold Krug corn as was intended, he selected
388 ears as a foundation and began inbreeding them in 1925. Within four
years Pfister reduced the genetic variation in Krug corn to only four inbred
lines, which he then used to produce hybrid corn. In 1934 his single-cross
hybrid produced the top yields at the University of Illinois trial plots (fig.
41).[46]

Pfister's success had been foreshadowed by the work of the two leading
figures in hybrid corn's early history, Henry A. Wallace and Eugene D. Funk.
Both Wallace and Funk had seed corn businesses which, like all others until
the mid-1920s, were based on developing new varietal crosses from the existing Corn Belt Dent varieties. The germ plasm necessary for this effort had
been slowly migrating west as the Corn Belt grew. The Reid's Yellow Dent
that Wallace marketed (and that he suggested started with a single Flint-Dent
cross made by Robert Reid) was grown in Iowa, had ostensibly emerged in
the Illinois Valley, and had been derived from seed brought from southern
Ohio. Leaming, the oldest named Corn Belt Dent variety, originated in
southwestern Ohio in the 1820s. A more uniform, 120-day variety was selected
from Leaming planted at Champaign, Illinois, in the 1880s. Subsequent
breeding near Galesburg, Illinois, produced a Leaming variety with deeper

kernels.[47] The Funks marketed Funk's Yellow Dent, which they developed from Reid's Yellow Dent. Further crosses made with the Leaming line produced still other new varieties.

Interest in the development of new types of corn began to emerge in the 1890s, at least a decade before Shull advanced the theoretical arguments that would eventually lead to successful hybrid corn production. The stimulus in the 1890s was not science, but rather the awareness that corn production would no longer increase simply by the expansion of the Corn Belt. On a national basis, corn yields did not appreciably increase from the 1860s until the adoption of hybrid corn became widespread in 1940. The size of the corn crop depended mainly, apart from the weather, on the number of acres in production. Expansion of production had come largely from the expansion of the Corn Belt itself. Given the declining harvests that characterized the southern part of the region, the lack of alternatives to the west, and the comparatively slow rate of advance to the north, there was no longer reason to anticipate that production would appreciably increase. That, in turn, increased the value of lands where corn yields were the largest. Corn Belt farmers found it necessary to derive more income per acre from fields that were increasing in value, and their only means of doing so was to increase corn yields.

Eugene Funk was one of many grandsons of Isaac and Cassandra Funk, who had settled in McLean County, Illinois, in the 1820s. By the latter half of the nineteenth century the Funk brothers, cousins, and grandchildren owned more than twenty thousand acres of land in McLean County (fig. 42). Much of it was in a contiguous block southwest of Bloomington, and nearly all of it was excellent for raising corn.[48] Eugene attended Yale but cut short his education for the chance to visit Europe, especially to observe farming methods. When he married in 1894, he interrupted his honeymoon to visit with the president of Northrup-King Seed Company in Minneapolis. He organized Funk Bros. Seed Company in Bloomington in 1901 and began experimenting not only with corn varieties, but also with soybeans, clover, alfalfa, and improved livestock nutrition.[49]

The Funks' acreage was large enough to permit wide separation of the varieties they were growing and thus to ensure that each could remain pure. What Shull had demonstrated in 1908 was that any open-pollinated maize variety, which consists of a range of heterogeneous genotypes, could be reduced in five to seven generations of selfing to a pure line, or inbred. This was to be achieved by selfing individual plants. But the Funks' isolated fields apparently achieved in a somewhat longer time the nearly identical result. The theory that explained this outcome was brought by James R. Holbert, a Purdue University agriculture graduate who had gained experience in the Shull-East methods and was hired by Eugene Funk in 1913.

Fig. 42. Lafayette Funk homestead at Shirley, Illinois, part of the large Funk holdings in southwestern McLean County.

As Holbert later explained, when he came to work for Funk, he "started with their [East's and Shull's] principles, but not their germ plasm."[50] Hybrid corn for the Corn Belt would have to be derived from varieties already growing there, a limitation demonstrated in 1915 when some of the seed from East's experiments in Connecticut was grown without success in Nebraska. Inbred lines were developed from the many Corn Belt Dents that had desirable traits known to breeders who had been desperately trying to improve them for years. Wallace developed his own inbred lines and produced a hybrid, Copper Cross, which won him a gold medal in the 1924 Iowa Corn Yield Contest. Funk's first hybrid appeared in 1922, and he began advertising hybrid seed corn in his catalog of 1926, the same year that Wallace formed the Hi-Bred Corn Company.[51] Wallace's operations around Des Moines and the Funk farms in McLean County became the two poles of innovation in the early years of hybrid corn.

The first cross of two inbred lines restored vigor and usually increased yields beyond those of the open-pollinated varieties from which the inbreds have been derived through selfing. This was impractical in terms of seed production, because inbred ears are small and typically contain a greatly reduced number of kernels, making the production of hybrid seed a slow process. Between 1915 and 1918 Donald Jones, working at the Connecticut station, had demonstrated that the problem of low seed production could be overcome by making double-cross hybrids.[52] Jones found that by crossing two single-crosses, each of which had well-filled ears, hybrid seed production was markedly increased. Although a reduction of 15% to 20% in yield resulted from the use of double-cross seed, the greater productivity of seed grain made this the favored production method until the 1960s.

Because hybrids perform well one year only, farmers had to be taught that it was necessary for them to purchase seed each year. The practice of saving the best-looking seed from the best plants was no longer to be followed, nor even was any corn to be saved as seed. State extension agents continued to instruct farmers through the 1920s on various labor-intensive means for carefully selecting seed, an activity that sometimes had the effect of retarding adoption of the new hybrids. As Deborah Fitzgerald has written, "hybrid corn rendered these hard-won lessons pointless and even of negative value, for many of the farmers who decided to try the new hybrids persisted in the traditional methods of growing corn."[53]

Despite the inertia of old practices, Corn Belt farmers quickly adopted the new hybrids once their higher yields were known. Zvi Griliches used annual data on hybrid corn acreage recorded in dozens of crop reporting districts to model the decision-making process involved in hybrid-corn adoption and diffusion.[54] I solved his equations for 10% and 50% adoption dates to produce diffusion maps (figs. 43, 44). The 10% adoption map (the year in which 10% of corn acreage was planted in hybrids) shows an axis of early trials between the Wallace territory in Iowa and Funk's McLean County base. The 10% threshold was reached earlier to the north of this axis than it was to the south, a reflection of the optimism of new producers versus pessimism among those who had already diversified away from intensive livestock feeding.

The 50% adoption map is similar although not identical (fig. 44). Once an innovation is introduced, its rate of spread depends largely on local attitudes and the relative amount of positive or negative feedback existing adopters give to those who have not yet made the shift.[55] The 50% map, superimposed on the map of corn production in 1939, shows that hybrid corn was adopted rapidly in those parts of the Corn Belt where production was already most intensive. Illinois's corn crop increased nearly 40% between 1929 and 1939, Iowa's by 20%, mostly because of higher yields rather than acreage increases.

Hybrid corn encountered relatively greater resistance along the western edge of the Corn Belt, where droughts in the 1930s evaporated farmers' interests in corn production. Eastern Nebraska reached the 50% threshold two years later than western Iowa. Kansas showed little interest in hybrids. Missouri, which had been slowly receding from the Corn Belt, lagged behind

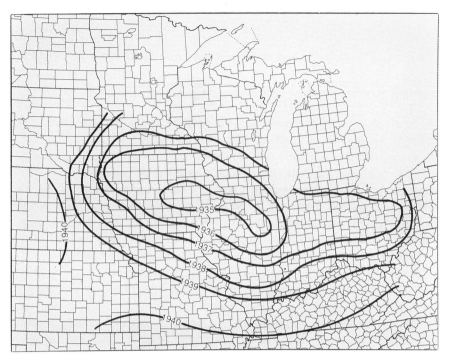

Fig. 43. Ten percent hybrid corn adoption, 1930s, based on Griliches's logistic equation (see chap. 10, note 54).

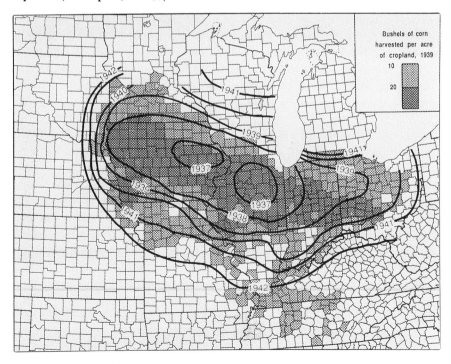

Fig. 44. Year of 50% hybrid corn adoption superimposed on map of corn production in 1939.

the other major corn states in the rate of adoption but nevertheless doubled its corn production between 1929 and 1939. On the northern fringe, Wisconsin and Minnesota farmers apparently were convinced more easily than those in Michigan. To some extent, hybrid corn's adoption mirrored the late 1920s, when Krug corn spread rapidly through the heart of the Corn Belt. Griliches concluded that hybrid corn was adopted earliest by those farmers who stood to profit the most from growing it.[56] That those same farmers had been quick to buy Krug corn ten years earlier reinforces the same point.

Hybrid corn's introduction had other impacts as well. Mechanical picking of ear corn was a problem with the older, open-pollinated varieties. Hybrids hastened the shift to greater mechanization, because the new corn stalks were sturdier, stood in the field longer in case the harvest was delayed, and were better able to engage the corn picker's knives without falling.[57] Corn hybrids made even better use of soil nutrients than the older varieties had, and farmers soon discovered that greater returns were obtained from the application of increased amounts of fertilizer, a development that anticipated the still larger machinery and costlier chemical fertilizer inputs that would come in the 1960s.

Better seed, stronger plants, higher yields, and more machinery all added to the corn surplus that government policymakers were already struggling to control in the 1930s. The first AAA, declared unconstitutional in 1936, was replaced by a second, enacted in 1938, that aimed to control farm surpluses rather than farm production as the first act had. The 1938 AAA included the concept of the "ever-normal granary," a recognition that surpluses were desirable, as they had been during the drought years of the 1930s, when grain stocks were depleted, but that the size of the surplus had to be checked. Production was to be limited under acreage controls announced by the Secretary of Agriculture, Henry A. Wallace. However, if farmers planted the hybrid seed corn that Wallace's Pioneer Hi-Bred Corn Company, among others, marketed, they could take advantage of government payments for acreage reduction and simultaneously harvest more bushels of corn on fewer acres.[58]

World War II brought a temporary end to the surpluses. In a move hailed by many Corn Belt farmers, the USDA released one hundred million acres of land back into production in 1943.[59] But a corn surplus returned soon after the war ended, acreage allotments were authorized for corn in the commercial growing areas in 1948, and production controls were imposed for the first time in 1950. Farmers had to comply in order to qualify for a government price-support loan for their crop. Corn acreage decreased nearly 20% under the program.[60]

Corn surplus or not, the trend toward a smaller Corn Belt was reversed by the new hybrids, and the region began to expand once again in the late 1940s. Although the greatest intensity of production remained in the swath from southwestern Ohio to western Iowa, which had been the heart of the Corn Belt for decades, production was gradually shifting northward. New hybrids that performed well under conditions of a short season and long day-lengths stimulated the existing northward movement. Irrigation allowed corn production to return to east-central Nebraska by the late 1940s, while new schemes in the central Great Plains would soon create new oases of corn production where even wheat crops had often been marginal in earlier years. Engineering a crop to suit the environment in which it is grown, tailoring it also to fit the machinery used in the planting and harvest, became commonplace by the 1950s. Technology was changing the farm and as a result technology also began to reshape the Corn Belt region.

11

West to the Plains

RAILROAD TRANSPORTATION OF live animals and refrigerated shipment by rail of dressed meat were the two practices that enabled Chicago's packers and stockyards to attain preeminence in the nation's meat trade by the 1880s. St. Louis, Cincinnati, Milwaukee, and Louisville were well-established packing centers before Chicago gained control, but they became subordinate in a transportation system based on railroads. Chicago's Union Stock Yards became the model for livestock marketing in the railroad era. Similar facilities were constructed at Kansas City (1871), East St. Louis (1872), South Omaha (1884), and South St. Paul (1886), forming a semicircular ring of livestock assembly points west of Chicago.

Between 1887 and 1910 a still more distant line of stockyards was developed at Wichita, Fort Worth, Sioux City, St. Joseph, and Oklahoma City.[1] All of these stockyards served local meat and by-products industries owned by the major packers, but they also supplied live cattle and hogs to the Chicago market. The unrivaled livestock market and meat-packing center of the nation even through the first half of the twentieth century remained at Chicago, the focus of a railway web connecting farmers, packers, and markets. Chicago's position was defined in terms of locational advantages that gradually disappeared, however, and the city eventually lost both its packinghouses and the companies that owned them.

Counting the total numbers of cattle, calves, hogs, and sheep, over one billion meat animals—worth nearly forty billion dollars—passed through the Chicago Union Stock Yards during its 105-year history. Roughly 20% of the hogs and 30% of the cattle were reshipped from Chicago, mainly to eastern cities. The rest were processed in the packinghouses that lined Union Stock Yards. Chicago's packers processed nine hundred million hams and a quarter of a billion sides of beef in their century-long history.[2] By the 1950s, after Chicago's packinghouses had become antiquated, their owners began to phase out operations in the city, and in 1970 the great Union Stock Yards closed permanently. That such a business could be built, operate for over a century, and then vanish without a trace confounds attempts to explain the concentration of meat packing at Chicago in terms of "natural advantages" the city supposedly possessed.

Gargantuan stockyards like Chicago's were created in an age when railroads were the principal means of transporting live animals to market. Late nineteenth-century Chicago was never as close to the center of hog production as Cincinnati had been in earlier years, but Chicago's rail access to all parts of the Corn Belt overcame its handicap of peripheral location. Railroads were ideally suited to mass a product at a single destination and were thus an important factor in concentrating the industry. Scale and agglomeration economies resulting from geographical concentration made further profits possible through vertical integration. The manufacture of edible by-products (shortening, gelatine) as well as the inedible (hides, soap, glue, grease, fertilizer) was most profitable at locations that handled a large volume of animals. In 1912, Swift and Company presented statistics claiming their meat sales produced insufficient revenue to cover the cost of the livestock purchased; it was the sale of animal by-products that made the whole operation profitable.[3]

Chicago was the industry prototype, Kansas City its first copy. All of the "Big Five" packers (Armour, Swift, Wilson, Cudahy, and Morris) were represented in both cities. Major public stockyards were corporations in which the major packers generally held a controlling interest. When South Omaha's packing industry was organized in the 1880s, Armour, Cudahy, and Swift built plants there, and each received part ownership of the yards in exchange. Swift and Morris gained control at East St. Louis in a similar fashion. The Federal Trade Commission's investigation of the packing industry in 1917 revealed that the Big Five dominated the dozen largest packing centers in the nation.[4] Armour and Swift each had plants adjacent to the stockyards in the nine largest packing centers: Chicago, Kansas City, South Omaha, East St. Louis, St. Joseph, Fort Worth, South St. Paul, Sioux City, and Denver (fig. 45).

Chicago and its many scaled-down copies, through collusion of the major packers, dominated the industry in this manner until 1920. In that year the U.S. Supreme Court decreed that the Big Five would sell their interests in the stockyards and would not operate retail stores (the latter a rather unsuccessful and short-lived attempt at further vertical integration). At the time of the consent decree, Chicago was by far the largest packing center, processing more hogs than St. Louis, Omaha, and Kansas City combined. Why the Chicago concentration continued—for decades after the Big Five had established plants nearer the sources of supply in Iowa and the Missouri Valley—is difficult to explain. Rate differentials between shipments of livestock versus dressed meat are insufficient to explain the size of operations at Chicago, and the city offered no advantage in terms of labor. Whatever led the Big Five to concentrate operations at Chicago, they eventually fell victim to their own strategy.

Many country interests, including those of the small meat packer and the livestock feeder, were opposed to the concentration at Chicago. An early sign of change came in 1891 when George A. Hormel opened a successful meat-

Fig. 45. Former office building and packing plant of Armour and Company adjacent to the South St. Paul stockyards.

packing operation on the banks of the Cedar River in Austin, Minnesota. Hormel's location was peripheral to the Corn Belt at that time, and there was no public livestock market at hand where a large and dependable supply of animals might be purchased. But Hormel's factory, and others that soon appeared in Iowa, demonstrated that terminal stockyards were not a necessary part of the business.[5]

Changes in marketing took place as well. In the 1910s Corn Belt farmers organized cooperative livestock-marketing associations for the purpose of co-ordinating shipments and controlling the supply at terminal yards. They were not very successful, in part because the packers countered with direct buy-ing—sending company buyers to the country to purchase stock either at auc-tion or at local assembly points.[6] Direct buying was quickly seen as a boon to the hog farmer, attempts to control supply at the terminal stockyards were abandoned, and farmers focused their efforts on obtaining the highest price, whether from a traveling stock buyer or a local packer.

Motor-truck transportation first began diverting attention away from the major stockyards in the 1920s. Truckers could haul stock, say, from eastern Iowa to Chicago, but they typically charged a higher rate than the railroad did on such a distance. It was the shorter trips to more local packinghouses that showed truckers' advantage over rail. Despite their eagerness to attract

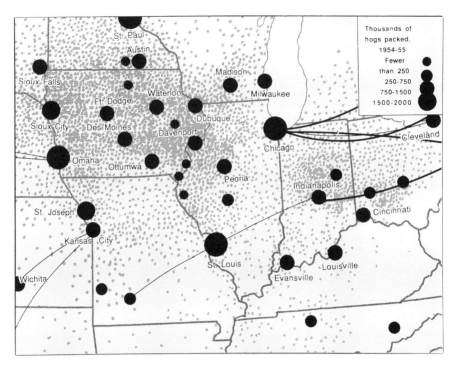

Fig. 46. Pork packing in 1954–1955. Each dot represents ten hogs sold for slaughter, 1954. Flow lines represent shipments of live hogs outside the Corn Belt (see chap. II, note 8).

truck business, terminal stockyards held an advantage only under the assurance of volume, rail transportation. And as more livestock were hauled to market in trucks—and hence more often to local packers—the terminal markets declined. Only one-third of the livestock arriving at Chicago in 1948 came by rail; truckers accounted for virtually all arrivals by the mid-1950s. Nationally, the proportion of cattle purchased in terminal markets fell from 91% to 46% between 1925 and 1960; the corresponding figures for hog marketing over the same period showed a decline from 76% to 30%.[7]

Omaha began a brief reign as the nation's largest packer of both cattle and hogs in the 1950s. Chicago had declined in importance and so, eventually, would Omaha, but it took several decades for the transition away from terminal markets to be complete. The newer pork-packing plants at that time were in northern Iowa and in southern Wisconsin and Minnesota, where the Corn Belt was both expanding and intensifying (fig. 46).[8] Several Iowa cities became important meat packers by the 1950s. The handicap of peripheral location became more obvious for St. Paul and St. Louis as well as Chicago: none of the three was in a major producing area, and the trunk-line railroads converging on them were no longer an advantage. Eastern stockyards, such as the one at Jersey City, which served the New York market, continued to receive a large volume of animals from the Corn Belt, in part because of a lingering preference in the East for "city dressed" meat. Rail shipment of live hogs and cattle to eastern markets thus continued, but Chicago's once-supreme role was diminished, in this case by the growing importance of Indianapolis. Omaha and Kansas City shipped hogs west and south to populated areas where swine production was inadequate to meet local demand. Chicago lost business in every direction.

In its heyday, the Corn Belt feedlot was an efficient system for fattening both cattle and hogs. Farmers raised corn, fed it to their cattle, and hogs devoured the waste. When stock were sold, they were shipped by rail to a distant packing center that processed pork, beef, and everything else it received. The all-purpose feedlot declined after World War II, because the geographies of pork and beef packing began to diverge as new, specialized packing firms entered the business. The tendency for pork and beef production to separate worked against the large packinghouse cities like Chicago, whose overbuilt factories required a large volume to operate profitably. Beef production was moving west, away from Chicago, even more rapidly than pork. In 1959 the ratio of cattle to hogs sold from farms showed a steady increase from east to west: .09 in Indiana, .18 in Illinois, .24 in Iowa, .63 in Nebraska, and 1.15 in Kansas.

Cattle feeding was moving west; hog production remained concentrated in the Middle West. Poultry, no longer "chicken and egg money" for Corn Belt farm families, was now a specialty itself, but of the South. The major packers tried simultaneously to diversify into Southern broilers, stimulate the demand for lamb, and whet the consumer's appetite for pork sausage, lean bacon, and juicy steaks. In contrast, the smaller specialty packers were taking advantage of the growing geographical separation between the various types of meat production by opening single-purpose production plants as near their sources of supply as possible.

One system was bound to erase the other. When the mighty Swift and Company closed its Chicago plant for good in 1959, they replaced it with a small beef processing facility at Rochelle, Illinois, a city of seven thousand people in the cornfields seventy-five miles west of Chicago. Within five years Swift built packing plants at Grand Island, Nebraska; Guymon, Oklahoma; and Clovis, New Mexico. Swift closed its Omaha plant in 1969. The company diversified into soybeans, Saskatchewan potash, and Florida phosphates; they moved into life insurance, and purchased two Oklahoma oil refiners. In 1972 Swift became the food products division of a company whose holdings included rental cars, ketchup, whiskey, gasoline additives, hosiery, and loudspeakers, but was pruned from the corporate tree in 1980.[9] Wilson and

Company was acquired by Ling, Temco Vought (LTV Corporation) in 1971 and was relocated to Oklahoma City but remained a meat packer and manufacturer of sporting goods. Armour changed to Armour-Dial and then was purchased by the Greyhound Corporation. In 1984 Armour became a subsidiary of ConAgra, an Omaha agribusiness conglomerate that started with poultry, eggs, and catfish in the South, had major holdings in Puerto Rico, and became a leading grain miller and feed producer.[10]

Three separate trends had converged to produce these chaotic changes in the industry Philip Armour and Gustavus Swift had been instrumental in creating. The stranglehold of the Big Five packers had been broken less by antitrust action than by the subsequent decline of the very stockyards/packing centers they had created. Trucks dispersed the business and helped the so-called "interior packers" thrive in the likes of Fort Dodge, Austin, Waterloo, and Madison. Stockyards had become outmoded. Hog production shifted west, but not nearly so far west as cattle feeding advanced after upland irrigation permitted large-scale feed-grain production in the Great Plains. The third influence came in the 1970s as a result of changes in the method of preparing meat for retail sale. All three trends reinforced the already evident westward movement. Beef packers moved the farthest west of all and eventually tugged the Corn Belt out into the High Plains.

Most people today are not surprised to learn that the beef sold in their local supermarket comes from Texas or Nebraska or Kansas. The picture of the Great Plains that many hold is one of herds of cattle grazing a grassy range stretching to the horizon. Slaughterhouses are not part of this picture, although many are now familiar with scenes of thousands of cattle confined in pens lined with feeding troughs, the contemporary feedlot. In fact, it has taken the better part a century to assemble all these components of the beef industry—grassy range, feedlot, and slaughterhouse—into a single locality. Until the 1960s the common method of supplying beef to eastern markets began with the time-honored, two-stage movement of young cattle from Wyoming or Nebraska to Corn Belt feedlots and from there to packing plants for slaughter.

In 1950 about 40% of Wyoming was in farm or ranch land that was unplowable or unsuitable for crops; another 38% was in federal grazing districts, national forests, or unappropriated public domain in which grazing was permitted. Most of the state was a range for cattle and sheep. Although sheep outnumbered cattle more than four to one, the state earned roughly equal amounts from the sale of each.[11] Hundreds of cattle feeders from Illinois and Iowa made annual trips to livestock sale rings in eastern Wyoming, as did others to the Nebraska Sand Hills, where they purchased young stock. Several hundred thousand cattle moved east from Wyoming each fall, destined eventually for feedlots throughout the Corn Belt.[12]

The annual movement of Wyoming beef cattle began to take new directions after World War II, in part a response to the changing geography of population. Wyoming cattle began to move south, to feedlots on the Colorado Piedmont, and also southwest, to California. In 1945 California ranked third among states in value of cattle received for feeding and breeding purposes, and by 1960 the state accounted for one-tenth of all cattle on feed in the United States.[13]

California's feedlots were concentrated in a seemingly unlikely place. One-fourth of the business was located in the Imperial Valley, an irrigated desert, below sea level in elevation, just north of the Mexican border. Cloudless summer days there see temperatures in the nineties or above, so hot that cattle feeders began constructing an artificial shade cover over a portion of their lots, thus making it possible to feed cattle the year around. Imperial Valley cattle tripled in numbers between 1940 and 1960.[14] Water diverted from the Colorado River supported large crops of alfalfa and hay grown on comparatively small acreages. Sugar beets, another important Imperial Valley crop, provided two sources of livestock feed—the beet tops, which are a desirable roughage, and the pulp remaining after sugar refining, a more concentrated feed that fills the role of grains in fattening. Although Wyoming and all the western states supplied cattle to California, Texas was the most important source of stockers for the Imperial Valley. Feedlots also appeared in southern Arizona in the 1950s, and they, too, relied mainly on Texas ranches for their supply of young cattle.

Here was a convincing demonstration that cattle feeding was not tied to a particular physical environment. The Corn Belt's summer rains, large corn crops, and lush pastures were totally absent in the Imperial Valley. Western cattle feeding was based on the drylot method whereby cattle are penned and fed a carefully constructed diet for rapid fattening. There were no hogs following cattle in the drylot. Manure produced by cattle (and hogs) became fertilizer for grain crops in the Corn Belt, but in the drylot manure simply poses an enormous problem in waste disposal.[15] Commercial fertilizers were cheaper as well as easier to apply, and many drylots devoted little effort to crops anyway, since it was cheaper to purchase both feed and fodder. Cattle gained weight rapidly on a concentrated diet of grains, but beet pulp and cottonseed hulls provided an adequate substitute; and both were cheaper, since they were waste products from local refineries or oil mills.

Texas ranchers who supplied young, grass-fed animals to the southwestern feedlots became interested in knowing more about this business, and in 1962–1963 a group of them toured several large lots in California and Arizona. According to Garry L. Nall, the most important discovery of the trip came at Casa Grande, Arizona, where they observed cattle purchased at the Amarillo Livestock Auction being fattened on milo shipped in from the High

Fig. 47. Custom feeding of cattle owned by absentees became a common practice in the southern Plains in the 1970s. Wheeler County, Texas.

Plains. "They were then informed that the fattened animals would be returned to Texas for slaughter."[16] Although Nall overstates the Texans' lack of familiarity with feedlots (the Llano Estacado of Texas and New Mexico had seen more than five dozen feedlots emerge by that time), most Texans were bystanders before 1960.[17] Between then and 1964, however, another eighty-four lots were constructed, increasing the total capacity of Llano feedlots to 352,000 head. The small-scale "farmer-feeder" of the early years was supplanted in terms of production by much larger custom cattle feedlots that did nothing but feed the cattle brought to them (fig. 47).

Agricultural settlers had come to the Texas Panhandle rather late, primarily after 1900, when large cattle ranches were being subdivided and sold as farms.[18] Southern ranchers and cash-crop farmers were prominent among the early arrivals. Irrigated cotton was established early, although dry farming of winter wheat and grain sorghum was the major activity. Corn Belt farmers also moved to the Llano Estacado of Texas after 1900 and have been credited with establishing most of the early feedlots there.[19] Cattle were fattened in much the same way as in Iowa, except that Texans and others who fattened cattle where corn had been proven a failure relied primarily on grain sorghums for their feed.

Sorghum, *Sorghum bicolor* (L.) Moench, originated in tropical latitudes, like maize. But unlike maize the various races of sorghum, including milo, kafir, feterita, and durra, originated in the savannas and semideserts of northern Africa.[20] They were taken from there to the subtropics of South Asia and to the Caribbean. Guinea corn was introduced to the United States at the time of the slave trade. In the 1850s interest was focused on sorgo (Chinese sugar cane, of African origin) for the production of sugar, syrup, and alcohol. Sixteen cultivars of sorgo were brought to the United States from Natal in 1857 by Leonard Wray, the English planter.[21] Although the use of sorghums for sugar never became practical, it was discovered that they produced grain under the same hot, dry conditions that caused corn to wither. Milo and kafir corn were brought to Texas by immigrants from Georgia and South Carolina around 1880.[22] The sorghum hybrids grown in the Great Plains today were derived from these introductions, although the plant has been greatly modified.

The several African races of sorghum were isolated and then crossed using the same principles involved in creating corn hybrids. Male and female flowers are found on different parts of the corn plant, which makes hybridizing experiments easy. Hybrid vigor was noticed in sorghum in 1925, but the development of hybrids was a slow process, because in sorghum both male and female organs occur in the same spikelet. It took thirty years of inbreeding to produce the male-sterile stocks necessary for hybrid seed production, although rapid adoption took place once the hybrids were introduced. By 1960 almost all of the sorghum acreage in the United States was planted with hybrid seed. Sorghums originally produced tall stalks and required hand harvesting; they were transformed to their present short stalks to permit combining.[23] Sorghum was originally a dryland crop, attractive because it was an alternative to corn, but the largest increase in production came when the new hybrid was grown under irrigation.

Texas accounted for nearly half of the sorghum grown in the United States in 1929. Production was concentrated in the Panhandle, especially near Littlefield, where a few ranches had become many farms. Kansas ranked second in grain sorghum production that year. Most Kansas farmers raised the crop, but only four counties, in the southwestern corner of the state, produced more sorghum than corn. The Dust Bowl years seem to have reinforced the drought-resistant qualities of grain sorghums in the minds of many Great Plains farmers, and by 1959 sorghum had become the major feed grain in most of Kansas and southern Nebraska. But for several reasons the crop never became a serious challenge to corn east of the Missouri River.[24] Most grain sorghums are too difficult for cattle to chew and require processing; they do not contain the Vitamin A cattle need, but corn and alfalfa do; and sorghums typically yield less grain per acre than does corn, when moisture is adequate for the latter. Although sorghums require less water than corn, cheaply obtained irrigation water erases the difference. It was the low price of irrigation that brought corn to the Great Plains.[25]

Western drylot cattle feeding emerged as much from local traditions of ranching as it did from crop farming. Most drylot feeders in the West were not simultaneously grain farmers as they would have been in the Corn Belt. Thus, cattle feeding first appeared in the Colorado Piedmont during a boom in sugar beets, around 1930. Water diverted from many small streams descending the eastern slope of the Rockies provided irrigation water for beets. Ranchers began raising beets and feeding the tops and waste to cattle. Wet beet pulp, which feeders obtained from the refinery processing their beets, was the foundation of Colorado Piedmont cattle feeding.[26] Water pollution problems eventually led to the discontinuance of wet pulp production, but dried beet pulp remained as a source of cattle feed wherever sugar refineries operated in eastern Colorado.

Agricultural colonists had begun irrigation near Greeley, Colorado, in the 1870s by diverting water from the Cache la Poudre River, a tributary of the South Platte. In the 1950s the Colorado-Big Thompson project was completed to capture water in the upper tributaries of the Colorado River and send it through the Rocky Mountains by tunnel to irrigate the South Platte River Valley around Greeley and scattered areas downstream from Fort Morgan.[27] Diversion from the Colorado River substantially increased the area under irrigation, creating a cropland oasis in a region that normally receives less than 16" of rainfall per year. Although corn was not a major crop there at the time, hybrid varieties were developed to withstand the cool nights typical of the cloudless High Plains.[28] In 1959 Morgan County, Colorado, produced over three million bushels of irrigated corn for local feedlots. Morgan County, joined by Scottsbluff County along the North Platte River in Nebraska (where irrigated beets similarly had given rise to cattle feeding) were notable western outliers of corn production that year and a hint of what was to come.

Corn was not a development that many people in the Great Plains anticipated, given what the Depression years had witnessed. Four of the ten driest years in Nebraska and Kansas in half a century occurred during the 1930s, when the two states produced less corn than Kentucky and Tennessee. The timing seemed wrong, but Roswell Garst, who divided his hybrid corn sales territory in Iowa with Wallace's Pioneer Hy-Bred Company, traded three Iowa counties back to Wallace for the rights to sell hybrid seed corn to the drought-stricken farmers of Nebraska, Kansas, and Oklahoma. Garst attended to field trials and engaged salesmen in the three states. Oklahoma's farmers, far from the Corn Belt mainstream, had not yet purchased the mechanical corn planters used in the Middle West; they were still planting corn by hand when the new hybrid seed was introduced. One of Garst's salesmen even ventured into the Arkansas Valley of Colorado, where a few farmers were irrigating corn. When told of his salesman's whereabouts, Garst retorted, "What the hell is he doing in Colorado?" and ordered the man back to Kansas.[29]

Irrigated truck crops, perhaps even an irrigated cash crop like beets, were thought suited to the confined sections of irrigable land along the Platte, Loup, and Arkansas rivers, but planting corn made little sense where the water necessary to grow it cost money.[30] Irrigation gained a foothold in the central Plains on the basis of fairly scarce local water resources, and its first expansion came at the expense of the Colorado River. The great extent of irrigation at present relies mostly on underground aquifers. The first water projects were labor-intensive, and water consumption was a concern. But as the source of water became more remote and its means of arrival more difficult to see, large-scale water consumers paid less attention to its wise use.[31] Early irrigation was designed simply to augment the then-existing agricultural economy of dryland farming. Eventually irrigation became the means to ignore climate insofar as possible.

"High Plains," a label sometimes used in popular literature to refer to the entire Great Plains, is actually a geologically distinct subregion (fig. 48). It is the remnant of an eastward-sloping plain that once extended beyond the present Missouri River. Its sediments were deposited by aggrading streams originating in the Rocky Mountains during the last forty million years. About three-fourths of the High Plains is covered by the newest of these depositional layers, the Ogallala formation, which is comprised largely of sand and gravel stream deposits.[32] Water is easily stored in the Ogallala aquifer, which, in places, has a saturated thickness of six hundred feet. Erosion and uplift of the land surface since the High Plains was formed severed the aquifer's connection to the Rocky Mountains, and thus recharge from the west is no longer possible. Erosion also caused the High Plains to retreat from its eastern margin, although surface deposits in parts of eastern Nebraska and central Kansas are in hydraulic connection with the High Plains/Ogallala formation and thus enjoy the same benefits of irrigation water. The High Plains aquifer is a massive sponge of groundwater that currently has some 170,000 operating wells penetrating it, most of them drilled for agricultural purposes in the past four decades.

Government policy hastened the adoption of pump irrigation. Because of acreage restrictions on wheat in federal farm programs, many High Plains dry farmers were forced to find an alternative source of income in the 1950s. Beef prices were attractive at that time, and a favorable loan policy made it possible to invest in wells, pumps, and movable sprinkler systems for irrigation.[33] Center-pivot irrigation machines, and others which can move over rolling topography, shifted attention away from alluvial valleys to the much

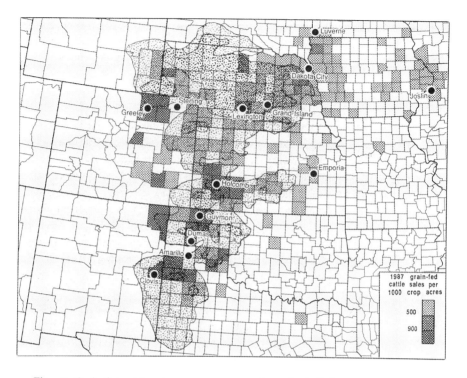

Fig. 48. Grain-fed cattle sales per 1,000 acres of cropland, U.S. Census of Agriculture, 1987. The stipple pattern shows the High Plains aquifer in terms of a saturated thickness greater than (darker shade) or less than 60 meters. Town names indicate major centers of beef packing in recent years.

larger expanses of upland that only rarely have gentle slopes long enough to permit ditch water to flow freely. Center-pivot irrigation is less affected by topography and is oblivious to the legal battles over surface water rights, because it uses water pumped directly up from the underlying aquifer. By 1977, 3.5 million acres of land were being irrigated by center pivots in Nebraska, Texas, Colorado, and Kansas (fig. 49).[34] The High Plains aquifer supplied most of the water.

Pump irrigation was especially suited to the Llano Estacado, where streams flow far below the level of flat, cultivable uplands. Texas High Plains cattlemen had only 221,000 cattle on feed in 1966, but six years later the number had increased to 1.68 million, concentrated mostly in the triangle between Amarillo, Clovis, and Lubbock. By 1973 High Plains feedlots, from Colorado to Texas, finished 3.2 million cattle each year.[35] Corporations were attracted to the profits to be made in cattle feeding; their desire for still quicker turnover increased the amount of growth hormones administered to cattle for purposes of making a more rapid weight gain. Nonfarm investors

Fig. 49. Center-pivot irrigation of corn and soybeans in Platte County, Nebraska.

found that owning cattle, but having them custom-fed by a commercial feedlot, offered tax-deferral incentives. Custom feeding became more common by 1973, but in that year the beef business began to experience difficulties. Cattle prices were reduced by half between 1973 and 1975, increased grain exports from the United States drove up the price of feed grains, and hundreds of feedlots went into bankruptcy.

Texas cattle feeders had begun large-scale operations in the 1960s by feeding both processed cottonseed hulls and milo. Consumer preference for corn-fattened beef remained an obstacle until the 1960s, when a large feedlot near Lubbock, Texas, won a contract with a major national grocery chain to produce milo-fed cattle.[36] But once those High Plains grain farmers, who had formerly sold their sorghum to local feedlots, began selling it for a higher price on the export market at Gulf Coast ports, and thus satisfying part of the world's demand for feed grains, corn began to replace milo in the High Plains. In 1969 none of the thirty-six feed-grain-producing counties south of the Arkansas River produced more corn than sorghum; in 1987, twenty-one of forty-four feed-grain counties produced a larger amount of corn.[37]

An open pan can evaporate up to eighty inches of water a year in the Colorado High Plains, perhaps five times the annual precipitation.[38] Irrigation substantially increases green-plant growth in such an environment, where evapotranspiration demand greatly exceeds water supply. The limits posed by semiaridity can be overcome by irrigation, but the lack of precipitation also means that underground aquifers that irrigators depend on will recharge slowly. Reliable estimates have water seeping into the High Plains aquifer at less than one-third the rate it is being removed.[39] Much of the recharge occurs in the Nebraska Sand Hills, where the saturated layer is thickest but where crop acreage is small because the sand dunes cannot be cultivated. Little recharge takes place in parts of the Texas and Oklahoma High Plains, where water levels have already dropped more than one hundred feet and some irrigation schemes have been abandoned.

Regardless of these limits—which were well known before widespread pump irrigation began in the 1960s—the short-term benefits of irrigating corn, feeding it to cattle in lots, and preparing meat for sale all in the same locality have been enough to move more than half of the beef industry of the United States to the High Plains in the past twenty-five years. The feedlot concept itself is scarcely new, although corn shocks and hogs are no longer part of the scene. Rather, it is the availability of technology to mine fossil groundwater and sprinkle it over cornfields that has finally achieved closure in a system geographically separated ever since cattle drives from Grand Prairie grasslands to Scioto Valley feedlots began in the 1830s.

Packing meat in the Great Plains means that the product has to be hauled farther to eastern cities, however. From 1880 until the 1960s the technology used to reach distant markets was the refrigerated railroad car, in which chilled carcasses swung freely from hooks. Proper ventilation was a major concern, but efficient use of the space was less a problem since the packers owned the cars and earned money every time one of them was in service. The Big Five had constructed large meat warehouses in major cities for the purpose of eliminating wholesalers. By the 1960s, however, supermarket and hotel/restaurant chains had become the principal buyers of beef. They took delivery from refrigerated trucks loaded at the packing plant.

The shift to truck transportation forced a more efficient use of refrigerated space. Iowa Beef Packers (IBP), a small firm at Dennison, Iowa, expanded by purchasing the Fort Dodge Packing Company in 1961 and soon began promoting the idea of "boxed beef." In the old meatpacking system, about one-sixth of the carcass was inedible fat and bone that was removed and discarded by butchers at the point of sale.[40] IBP began boning and cutting the meat, wrapping it for supermarket delivery before the product left the packing plant. The intermediate warehouse was eliminated, more by-products could be recovered, the weight shipped to market was decreased, and the meat-cutting done in the factory increased the sale price.

IBP expanded by purchasing other packing firms. In 1965 they opened a new plant at Dakota City, Nebraska, which they claimed was "the most complete and efficient facility ever created for the slaughtering, breaking, fabricating, and boning of beef."[41] IBP overcame the opposition to boxed beef by butchers' unions, and their marketing was more aggressive yet. In 1974 Currier Holman, a co-founder of Iowa Beef, was convicted of conspiring with New York organized crime figures in an attempt to bribe IBP's way into the city's meat business.[42] Federal Trade Commission and Occupational Safety and Health Administration inquiries into IBP and the meat industry in general have been frequent since the early 1980s.

In their survey of business strategies in the packing industry published in 1973, Robert Aduddell and Louis Cain wrote, "vertical integration did not go backward into livestock production because of the relatively low profits in that field."[43] This was true before large-scale irrigation reached the High Plains, but not after. Warren Monfort of Greeley, Colorado, started one of the largest feedlots in the Piedmont in the 1940s. By the mid-1960s Monfort was producing twenty thousand head of cattle for slaughter annually. The next step was entry into the packing business. Kenneth Monfort combined range, feedlot, and packinghouse into one company whose main feedlot near Greeley, often cited as the world's largest, was equipped to feed two hundred thousand head of cattle in the late 1970s. Monfort's Greeley packing plant, plus the former Swift plant at Grand Island, Nebraska, that Monfort purchased in 1979, could process forty-six hundred head of cattle per day.[44] Like IBP, Monfort specialized in boxed beef.

Monfort's cattle were corn fattened, but corn production near Greeley was insufficient to supply the feedlot. Warren Monfort had solved this problem in the 1960s by shipping trainloads of irrigated corn from Cozad, Nebraska, to Greeley. By 1982 eastern Colorado grew enough irrigated corn to supply local needs, and the imports from Nebraska were discontinued.[45] By then, cornfields stretched up the North Platte Valley into Wyoming.

If the early 1960s was the high point of flexibility and specialization in the meat industry, the 1980s brought a return of concentration and a renewed interest in vertical integration. Monfort diversified into lamb, built one plant in Nebraska to cook meat for frozen entrees, and constructed another in Jacksonville, Florida, for the single purpose of forming hamburger patties. In 1987 Monfort was swallowed by ConAgra of Omaha.[46] IBP became the agriculture division of Occidental Petroleum Corporation in 1981. The next year they diversified into pork and are now a major hide tanner. IBP currently slaughters eight million cattle a year and operates eleven processing plants from Texas to Illinois to Washington.[47] More than half of the cattle IBP slaughters are fed grain irrigated from the High Plains aquifer. IBP is the world's largest beef producer and accounts for approximately one-fourth of United States production. IBP's share of the pork trade is smaller, but growing.

The Big Five packers' abysmal record of labor relations in years past was echoed in the new western beef industry almost from the start. Workers struck Monfort's Greeley plant in 1979, prelude to a two-year total shutdown of the facility. Iowa Beef's Dakota City plant was the focus of strikes and violence a decade earlier. The company had no contracts with organized labor for many years but now reluctantly admits to having agreements at its Amarillo, Texas, and Pasco, Washington, beef plants. Jimmy M. Skaggs has suggested that, although IBP has typically paid high wages to its employees, the company deliberately located in states like Iowa, Kansas, Nebraska, and Texas, where organized labor is comparatively weak.[48] Such an interpretation gives insufficient weight to other factors, but the desire to avoid organized labor contracts remains evident in IBP's public statements. The enervating job of slaughter, personified in the Lithuanian immigrant, Jurgis Rudkus, in Upton Sinclair's, *The Jungle*, still remains work for some of society's least established members, just as it was in Sinclair's day.[49]

Slightly more than half the nation's value of beef shipments now originates in Kansas, Nebraska, and Texas (fig. 48). There has been a westward movement within those states as well: plants in Omaha, Wichita, and Fort Worth have closed, replaced by still larger ones in the likes of Lexington, Nebraska; Holcomb, Kansas; and Dumas, Texas. As Michael J. Broadway and Terry Ward have written, "the U.S. meatpacking industry has been transformed from an urban to a rural-based industry."[50] Today's packing plants are built in the country, and they employ workers who often commute many miles to the job. On a regional scale, the outlines of the western Corn Belt and the western beef industry are nearly identical; both follow the outline of the High Plains aquifer, the lone significant exception being the irrigated South Platte Valley between Greeley and Fort Morgan, which is supplied water from the Colorado River, across the mountains. None of it would be possible without irrigation.

A reorganization of Corn Belt farming was stimulated by the movement of cattle feeding to the Great Plains. Cattle numbers declined most noticeably in the mid-Mississippi Valley, where cattle feeding had once been a major source of income. Farmers continued to raise hogs, but they plowed under the pastures that had been used for cattle and used them instead for corn and soybeans. An additional stimulus to cash-grain farming came in the early 1970s when the export markets for both corn and beans seemed to grow stronger every year. Logistics were inverted once more, and the Mississippi River again became the route to market, only this time for grain barges. In Illinois, Iowa, Minnesota, and Wisconsin 15.7 million new acres of corn and soybeans—more than 24,000 square miles of additional cropland—came into production between 1969 and 1978 as farmers, following the advice of the day, acquired as much land as they could work and planted their crops "fence-post to fence-post."

12

The Corn Business

OWNERSHIP OF THE means of production has been in the hands of those who have worked Corn Belt land—at least that has been the ideal. While farm-management decisions in corn/livestock agriculture probably are best made at the scale of the individual farm, landlords and their tenants have nonetheless been common at times.[1] The practice of renting farms entered the Corn Belt in the Virginia Military District and was taken west to the Grand Prairie in the 1850s. In 1880, when the first national statistics on tenancy were published, 52% of the farmers in Madison County, Ohio, were tenants; five Grand Prairie counties had at least 40% tenancy, but no other Corn Belt counties had tenancy rates as high at that time.

The unrivaled fertility of Corn Belt land attracted investors who bet its price would increase, and they, in turn, rented acreages to tenants whose cash rents paid interest on the land. A detailed analysis published in the census of 1910 showed that the average tenant farm in the Corn Belt was relatively more valuable in land than an owner-operated farm, but the tenant farm typically had a smaller investment in buildings and improvements. As a rule, tenants raised more corn than any other crop, they used the land more intensively, and they were more likely to sell grain for cash than feed it to stock on their farms.[2] Cash-corn farming was thus a typical response of the tenant who had to pay an annual rent. By the 1910–1925 period, tenancy dominated the Grand Prairie, and it was emerging as the common mode of tenure on the Des Moines lobe in Iowa (fig. 50).[3]

Rising land values accompanied an increase in tenancy on the wet prairies. Drainage had made these the most productive lands in the Corn Belt at a time when further geographical expansion of corn production remained in doubt. An even higher value was thus placed on drained prairies. While the cost of drainage itself added to the value, land prices spiraled during the World War I boom and continued an upward trend until the 1920s as investors bought and sold farms speculatively. Iowa corn land sold at an average of $82 per acre in 1910 but for $199 per acre in 1920; comparable figures for Illinois showed an increase from $95 to $164 during the same decade. The Corn Belt was, in the words of Theodore Saloutos and John D. Hicks, a "boom-stricken area."[4] Land became too expensive to own, and those who lost farms seemed ever less likely to return to ownership under the condition of rising land prices.

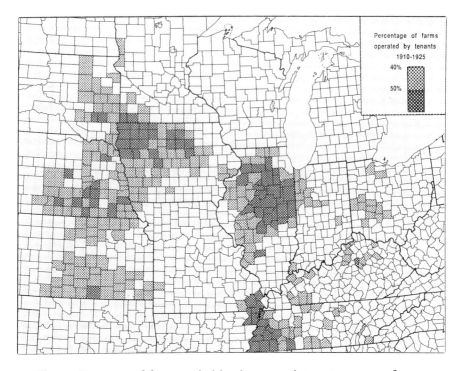

Fig. 50. Percentage of farms worked by share or cash tenants, average of 1910, 1920, and 1925 censuses (U.S. Census of Agriculture).

By the 1930s farm tenancy was more prevalent in the heart of the corn and wheat belts than it was anywhere else in the United States outside the Cotton South. Unlike cotton sharecroppers, however, Corn Belt tenants were predominantly of the cash-rent variety. Some were formerly farm owners and had been forced into tenancy as a result of mortgage foreclosures; others were classified nonowners because they worked farms belonging to relatives. Regardless of the cause for tenant status, the federal government viewed the steady increase in tenancy with alarm. In 1935 tenants outnumbered full owners in over 60% of Illinois's counties and in nearly 80% of Iowa's (fig. 51). Tenancy was widespread in the Corn Belt; it similarly dominated the wheat-raising areas of the Great Plains; yet the dairy-specialty areas of Michigan, Wisconsin, and Minnesota had comparatively few tenants. Even within the Corn Belt there were marked differences: tenancy had started to decline in Ohio and Indiana by the 1930s, even though it was on the rise in Nebraska and Kansas.

Changing demographics was one cause of geographical variation in tenancy within the Corn Belt. The generation of farmers who pioneered new

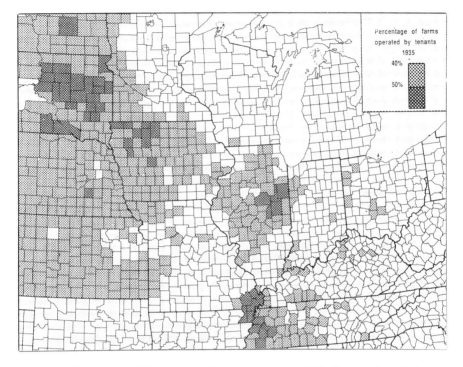

Fig. 51. Percentage of farms worked by tenants in 1935 (U.S. Census of Agriculture).

lands in the 1850s had retired by 1900; homesteaders of the 1860s similarly had been replaced by younger farmers by 1910. A wave of farm turnovers produced by this aging effect spread from east to west, lagging about one to two generations behind the early settlement frontier. Farms often were kept in the families of the original owner, but the land and buildings were rented. The practice of improving land and then selling it to move west was common in the nineteenth century. Tenancy was increased, because each such sale provided opportunities for nonfarming investors.[5]

But it was the wave of farm foreclosures sweeping the Corn Belt during the 1930s that most concerned both the farmer and the federal government.[6] Agricultural depression in the 1920s and the general depression that followed brought thousands of foreclosures, especially on the better lands where prices had advanced beyond what the land could be made to pay. Although tenancy sometimes is portrayed as a syndrome involving poor farmers on poor land, tenancy in the Middle West was unmistakably concentrated on the best land, especially prairies. Estimates of farm production costs made in the 1920s showed that the interest on land was the largest single item in the cost of producing a bushel of corn.[7] If interest on land was figured at 6% per year and typical rents were paid by tenants, the farmers who owned the land they worked sometimes were actually at a disadvantage compared with tenants.

Renting a farm from a distant landlord entailed some restrictions on how the land was to be used, but it also could mean greater profits and entail fewer risks. Any positive assessment of the benefits of tenancy has to be balanced against the negative side—the bad loans, foreclosures, and bank failures that eventually removed land ownership from the hands of local people and placed it in the portfolios of distant creditors. By 1932 insurance companies found that they owned about 20% of the farmland in Iowa. Roswell Garst sold a group of Iowa insurance executives on the idea of sharing the cost of hybrid seed with their many tenants, resulting in sales for Garst and greater income on farms that the insurance companies had come to own.[8]

Corn Belt farm tenancy continued into the 1930s, but the percentage of farms worked by tenants had reached a maximum by the time New Deal agricultural policies designed to restore ownership came into play. The practice of renting farms to tenants had helped perpetuate the old farm units intact. Tenancy had been sustained in part because, for many years, the lack of scale economies had kept Corn Belt farms of modest size. The average size of farm in Ohio, Indiana, Illinois, and Iowa was no larger in the 1930s than it had been at the beginning of the Civil War. The increase in farm mechanization—which made labor more productive—and the increase in size of farms also caused tenancy to decline. Increased mechanization made it possible to work more land, and thus farmers, whether owners or operators, sought larger acreages. Farms began to increase in size after 1940. There was only one tractor for every four farms in the Corn Belt in 1930, but by 1950 the number of tractors was equal to the number of farms.[9]

Farms decreased in number and increased in average size, and more tenants became owners. Part owners (farmers who owned only part of the land they worked) steadily replaced tenants from the 1940s onward. Owning part of the land and renting the rest on a cash or share basis became the most common mode of land tenure in the Corn Belt by the mid-1970s. Today part-owners own more equipment and work an average of more than four times as much land as do farmers who cultivate only the land that they own. Tenant farms still are common on the Grand Prairie and the Des Moines lobe, but high levels of farm tenancy turned out to be only a temporary condition in much of the Corn Belt. Farm mechanization and other technological changes in agriculture have been as responsible as any set of factors in reducing farm tenancy.

Land values had increased across the Corn Belt after 1900 as a part of nationwide prosperity and also because of the increase in demand for agricultural products. While corn's overwhelmingly most important use was for feed, industrial uses for corn did emerge slowly over the latter half of the nineteenth century. Starch accounts for approximately 80% of the kernel's dry weight, and thus corn became a substitute for wheat in the production of food and laundry starch, the manufacture of which began in New York and New Jersey in the 1840s. Corn was shipped there from Chicago via the Great Lakes ports of Buffalo or Oswego and then sent down the Erie Canal. Conversion into syrups and sugars began in the 1870s. Corn wet-milling (starch and glucose manufacture) did not become established on a large scale in the Corn Belt until 1879–1880, when plants opened in Peoria and Chicago.[10]

Corn milling enjoyed several periods of growth and expansion that were followed by consolidation of ownership and disappearance of smaller firms. The Corn Products Company, formed in 1906, operated a sizable plant in Pekin, Illinois, and opened a still larger facility at Argo, west of Chicago, in 1913. Augustus Eugene Staley, who would later become the first major processor of soybeans, entered the cornstarch business in Decatur, Illinois, in 1909. These two firms, plus American Maize Products Company at Roby, Indiana, near Chicago; the Clinton Corn Products Company at Clinton, Iowa; and Penick and Ford, Ltd., of Cedar Rapids, Iowa, accounted for 87% of the corn-grind capacity among U.S. manufacturers in 1958.[11]

Factories were located near the cash-corn sections of Illinois and Iowa, as would be expected. Government policy favored existing milling centers through application of the transit privilege on shipments of agricultural commodities, the same system of tariffs that governed soybean products. In-transit rates helped erase locational differences between corn millers. When a buyer for a milling company purchased a carload of corn on the cash-grain market at the Chicago Board of Trade, a freight contract on its shipment was also purchased. If the corn moved to a miller for manufacture into syrup, then the charge for hauling the corn to the milling center was normally subtracted from the charge for hauling the more expensive syrup to market, although with the addition of a small transit fee. Furthermore, since the corn lost about 40% of its weight in manufacture, the unshipped weight represented a transportation credit the miller could apply to the next shipment.[12] Millers understandably favored this system, which protected their investments in factories whose locations were no longer optimal. Transit privileges did not create the locational pattern of corn wet-milling plants but, because transit favored existing milling centers, it helped concentrate the industry at favored sites.

The 40% weight loss in corn wet-milling produced a transportation saving, although the weight-loss itself was avoidable. Unlike the economy of pork packing, where uses were found for nearly the entire animal, cornstarch manufacturers had no line of by-products. They used only about half the

Fig. 52. Truckers delivering shelled corn wait for an outbound tank-train of corn sweeteners to clear the crossing into the Archer-Daniels-Midland processing plant at Cedar Rapids.

weight of the raw corn and dumped the rest.[13] Starch manufacturers later learned how to recover the oil, and they also began using the dried residues to produce livestock feeds. The most important development in corn wet-milling came in the 1940s when commercially successful corn sweeteners were created. Eventually corn became an important source of sugar in the manufacture of many food products. Corn processing is now a major Corn Belt industry (fig. 52).

Introduction of a weak acid in the production of cornstarch converts the mixture to corn syrup or corn sugar, which is used in the manufacture of candy. A corn syrup of higher sugar content results from introducing an enzyme catalyst to the process. The A. E. Staley Company developed a dextrose-producing enzyme in the 1940s for the purpose of creating a sweeter, less viscous syrup. By the late 1950s the manufacture of ice cream, jams, soft drinks, and various prepared foods required more than one billion pounds of corn sweeteners annually.[14] With corn syrups established as an ingredient in the expanding array of factory-produced foods intended for direct human consumption, manufacturers of corn sweeteners intensified research programs to develop high-fructose corn syrups that had a cleaner, sweeter taste than the typically heavy syrups then available.

High-fructose corn syrup (HFCS) was marketed in the early 1970s and was adopted widely as a sweetener. It is produced by converting part of the dextrose into fructose, making a sugar equivalent to that produced from cane or beet. A second-generation HFCS, introduced in 1976, converted the existing 42% high fructose syrup to 55% HFCS, an even closer substitute for the traditional sources of sugar used in many foods.[15] Second-generation HFCS replaced sugar in soft drinks, salad dressings, frozen desserts, and cereals and also offered a food product that was lower in calories. By 1980, U.S. per capita consumption of HFCS was twenty pounds annually (dry basis). Perhaps the ultimate acceptance came in 1984–1985, when the three largest U.S. soft drink manufacturers approved 55% HFCS to be used for all sweeteners in their products, adding another billion pounds to the corn sweetener market.[16]

Each increase in knowledge of the uses of corn has led to further attempts to enhance its various qualities. Corn has been changed using methods ranging from simple selection to genetic engineering. A selection program begun at the Illinois Experiment Station in 1896 attempted to increase the desirable traits of high protein and high oil content. Seventy generations (seventy years) of selection produced corn that moved well beyond the foundation stock. The most extreme ear in the original population had an oil content 3.8 standard deviations beyond the average, representing a rare occurrence of about three ears in ten thousand. After seventy years the maximum oil concentration had been increased to twenty-seven standard deviations beyond the average in the initial population.[17] Efforts to create kernels with a still higher protein content for human consumption resulted in high-lysine corn, developed at Purdue University by Edwin T. Mertz and his students in 1963. Their "opaque-2" and "floury-2" differed from ordinary corn because they contained a single mutant gene that changed the protein composition of the corn kernel, nearly doubling the level of two important protein components, the amino acids lysine and tryptophan.[18]

Although gene splicing and the use of recombinant-DNA methods in biotechnology remain controversial with consumers of food, those same consumers tend to regard corn as environmentally "friendly," such as its use in the production of so-called biodegradable plastics. Cornstarch is an important ingredient in factory-produced foods because of its ability to hold flavors.[19] Starch made from corn has long been used as a sizing ingredient in the manufacture of paper and textiles. Industrial oils made from corn and soybeans are used in plastics, printing inks, caulks, and resins.

The sharp increase in gasoline prices that resulted from a reduced supply of foreign oil reaching the United States during the late 1970s helped focus attention on the use of corn to produce ethanol for blending with gasoline. "Gasohol," a trademark of the Nebraska Agricultural Products Industrial Utilization Committee, is defined as 90% unleaded gasoline and 10% fermentation ethanol.[20] In November 1978 the federal government reduced the excise tax on gasohol by four cents a gallon. Within one year most major gasoline refiners had built facilities for blending ethanol with gasoline, although interest declined after gasoline prices began to drop in the 1980s. Ethanol made from corn was foreseen then as a means of extending the gasoline supply. Its major use since the mid-1980s has been that of an octane booster, replacing the environmentally harmful tetraethyl lead in automobile fuels. Decatur, Peoria, and Cedar Rapids are now industrial energy producers, making both grain alcohol (neutral spirits) for remanufacture and ethanol for fuel. The federal subsidy on gasohol remains, to the delight of corn growers and processors.[21]

Several thousand years have elapsed since corn emerged as a food crop in the Americas. It has been grown as a feed grain for only a matter of centuries. Yet in the span of only a few decades corn, soybeans, and other oilseeds have become the principal raw materials used to produce a wide variety of biochemicals that are now accepted ingredients of prepared foods and are used to fashion many industrial products. Each transformation in the use of corn has required more intensive production and therefore has been accompanied by changes in the plant's culture. Indians planted the seeds of corn in hills, along with those of beans and squash. The hill was a miniature garden, bordered by others of like construction. The corn plant's tillering habit and extensive, shallow roots made hill culture, rather than row culture, the more desirable arrangement. This practice, which early Euro-Americans learned from the Indians, was continued into the nineteenth century, although the bean and squash crops often were omitted.

Nineteenth-century Corn Belt farmers refashioned the indigenous clustered or irregularly spaced hills into rows of hills, which could be plowed and cultivated in linear fashion. Sometimes ground was prepared with a lister, a double-moldboard plow used to throw earth up into parallel rows on either side of the furrow.[22] The field was listed in both directions, producing a square grid of rows whose intersections were "hills," spaced approximately four feet apart. Two to four kernels of corn were planted per hill. Farmers then cultivated to control weeds, following the listed rows in both directions, simultaneously mounding earth back upon the hills. The crop was laid by (cultivated for the last time) in July. A well-tilled and scrupulously weeded field produced twice as much corn as a similar one given only minimum effort.[23]

Mechanical corn planters and wheeled cultivators increased the amount of corn one farmer could produce, but they did not eliminate the need for cross-cultivation. Corn planters used through the 1950s produced a perfect geometric arrangement of corn rows in both directions, known as checkrows. The checkrow corn planter dropped seeds at the command of evenly spaced buttons on a wire that was stretched across the field and moved each time a traverse was completed. If three seeds were dropped at a time, given a germination rate of 70%, then nearly 80% of the hills would produce more than one stalk of corn. Together with the mechanical planter's usual 3.5' spacing between rows, the result was approximately eleven thousand plants per acre, producing a yield of sixty to eighty bushels, depending on the year.[24] Because much of the cornfield consisted of straight, open rows to permit cultivation, the land surface was subject to increased erosion. Checkrows made it impossible to plant along the contours of the land.

Some Corn Belt farmers used a four-field system of crop rotation to maintain topsoil and boost fertility. Those who followed the typical Pennsylvania practice began the cycle by planting corn on a plowed-under ley of clover or other leguminous hay that fixed nitrogen in the soil the year before corn was planted.[25] Corn was followed by oats and, the year following that, by winter wheat. Livestock manure was applied to the field in the years that corn and oats were planted. Soil fertility was restored the year following wheat by beginning the cycle again with a hay crop. Livestock were fed the hay and were turned loose to graze in the fall wheat fields, often in the oats and corn crops as well.

The four-field rotation incorporated animal wastes and green manure crops to restore soil fertility. It worked perfectly so long as equal acreages were devoted to each of the four crops each year, but the system broke down wherever farmers planted a larger corn acreage, as they did over most of the Corn Belt. Wheat was the crop usually omitted. Short-term tenants in the Corn Belt needed a cash crop and often could not count on working the same farm the next year. For both of these reasons they had less incentive to engage in crop rotation schemes. Those who owned the land they worked were more likely to rotate crops, but they also cut short the rotation to plant more corn. According to Wallace and Bressman's 1925 corn-growing textbook, 35% of the corn in the Corn Belt was planted on corn-stalk land, 45% on small-grain stubble, and only 20% on sod ground.[26]

Until the 1950s only two theories of corn production had ever been widely successful in obtaining larger yields. Both had a firm scientific base. From colonial times until the 1920s production was increased by making varietal crosses between the variable, open-pollinated Flints and Dents. The second theory led to hybrid corn and relied on the principle of inbreeding, or self-pollination, which reduced genetic variability. Hybrid vigor was captured to a maximum degree in the first cross that was made between inbreds. For purposes of increased seed production, the first commercial corn hybrids were double-crosses, derived from two single-cross parents. Male-sterile stocks were developed in the 1940s to make detasseling unnecessary. But even with these developments and the increased use of nitrogen fertilizers corn yields began to level off once again in the 1950s, just as they had in the years before hybrid corn was introduced, and just as they had in Woodford County, Illinois, by 1920. Double-cross hybrids had reached their own yield plateau.[27]

There was nothing like an equivalent of Mendelian genetics awaiting discovery in the 1950s that would increase corn yields as much as the early hybrids had. Corn breeders returned, instead, to the single cross. Limited production of single-cross seed had been the stimulus for creating double-cross hybrids in the 1920s and 1930s because single-cross ears were small and too imperfectly filled to be economical in seed production. Theoretically, however, the single-cross hybrid had twice the expected genetic gain of the double cross.[28] In the 1960s the solution was to make up for the inherent lower production of single-cross seed by increasing the application of herbicides, pesticides, and fertilizer. In so doing, seed output was forced upward to an acceptable level. There were no single crosses in production in 1960, but by 1980 85% of commercial corn came from single-cross hybrids.[29] Probably half of the increase in corn yields after 1960 resulted from planting the new seed, although seed production itself required larger energy inputs than before.

The balance of the increase in corn yield has come from a more intensive culture calling for increases in practically every input. Until the 1950s cornfields across the Middle West were little more than the ultimate, rectilinear-grid version of the hill-culture maize practiced by aboriginal North Americans. Space between rows was important because of the need to cultivate. Checkrows left too much land unproductive, however, and this distinctive, double-axis geometry of the cornfield soon disappeared, a casualty of the shift toward increased intensity. With herbicides there were fewer cultivations and cross-cultivation was eliminated entirely. Heavier use of nitrogen fertilizer allowed the rows to be spaced more closely, the plants set closer together within each row (fig. 53). Instead of twelve thousand plants per acre, the farmer could have twenty-five thousand if enough fertilizer was used, and yields upward of two hundred bushels could be expected on good corn land.[30]

The old corn pickers, which harvested ears of corn one or two rows at a time, were replaced by the grain combine, which sliced off eight rows, shelled the grain, and blew out the chopped stalks and cobs. Farmers ripped out old fence lines that had demarcated fields for generations, making new, elongated fields in order to reduce the number of times large machinery had to be turned at the ends of rows (fig. 54).[31] Larger fields worked by fewer people operating larger machinery usually meant an end to the old pattern of crop rotation. Continuous corn was possible as long as the soil's nutrients were replaced artificially each year. The corn plant was changed once again as well: easily snapped shanks were designed in the corn-picker era, but they

Fig. 53. Narrower rows and closer spacing of plants have doubled the amount of corn grown on an acre of land, a culture dependent both on herbicides and nitrogen fertilizer.

were replaced by sturdy shanks more suitable for the combine.[32] Since corn now came from the field already shelled, it had to be stored in an enclosed grain bin rather than in the old, air-ventilated corn crib. Some farmers purchased gas-fired grain dryers, which reduced corn's moisture content to bring a higher sale price.

Due to these many changes that took place through the 1960s and into the 1970s, it became increasingly easier for farmers to produce more corn, to store it on their farms until they were ready to sell it, and thereby to find greater interest in following the rapidly growing cash-grain markets. Food uses for corn, especially corn syrups, were expanding. Countries with rising standards of living were consuming more meat and hence had larger demands for feed grains. In 1971 President Richard Nixon rescinded the order requiring that half of the grain shipped to Communist countries move in U.S.-flag vessels.[33] In 1972 private traders sold grain to the U.S.S.R. and to China. The Soil Bank, which had diverted twenty-five million acres of land out of corn production beginning in 1956, was abolished in 1973. Thousands of farmers began to increase their acreages of corn and soybeans. Helping set the tone of the times, Secretary of Agriculture Earl Butz wrote, "corn, a golden grain, is as important to our wealth as gold itself."[34]

Fig. 54. Preparing corn land for planting, Boone County, Illinois. A herbicide tank is mounted on the high-horsepower tractor; ganged in tow are the applicators, disc harrow, and fertilizer tanks. The fence has been abandoned because the farmer no longer raises animals.

From 1972 through 1981 the United States exported 17.85 billion bushels of corn, more than triple the level of corn exports during the previous decade. The Soviet Union was a frequent customer, as were Japan, Korea, and Taiwan. Quantities of corn had long been sent abroad, but the U.S. grain trade until the early 1970s was oriented primarily toward wheat and was based on the practice of sending loaded ships to Europe. The St. Lawrence Seaway was a typical route to market. More than half of the grain-storage capacity among American terminal and port elevators in the early 1960s was concentrated at Great Lakes ports. Some of these elevators began storing less wheat but received more corn and soybeans.

New acreages appeared in the Great Lakes region, both within and outside traditional areas of production. Nearly four hundred thousand new acres of corn were planted on drained lands of the Saginaw Lobe and the "thumb" of Michigan between 1969 and 1978 (fig. 2). The Lake Maumee Plain, already a major soybean producer, increased its bean fields by three hundred thousand acres during the same period. Much of the increase in Indiana, Michigan, and Ohio was based on export demand. Although the St. Lawrence Seaway was the closest water route to the east, both corn and soybeans from the eastern Middle West also moved by rail to the port of Baltimore for export.[35]

A sudden expansion of corn acreage occurred in Wisconsin and Minnesota during the 1970s. Short-season hybrids made it possible to grow corn north to the limit of productive agricultural soils. More than five hundred thousand new acres of corn appeared where dairy farming had been the principal activity. While this development broadened the economic options of local farmers, it resulted primarily from the strength of the world market for corn. Grain barges were able to navigate the Mississippi River as far north as St. Paul. A farmer on the forest fringe one hundred miles north of there could be in the cash-corn business, because the cost of transportation to New Orleans was scarcely larger for the Minnesotan than it was for an Iowa farmer who also hauled corn one hundred miles to the riverbank. It was the same principle that, more than a century earlier, allowed meat packers along the Missouri River in northwest Missouri to compete in the New Orleans market with other packers who operated along the lower Wabash: distance up the river meant fairly little compared with distance to the river.

Although the Mississippi River had not been the Corn Belt's major highway to market for more than a century, the extremely low cost of barge transportation on the inland waterways led to a quick recapture of traffic as soon as the grain export boom got underway (fig. 55). Barges carried 51.6 million tons of grain in 1979, up from only 4.2 million tons in 1955.[36] The Illinois River contributed the most grain traffic per mile; dozens of small terminals along its banks received grain from trucks and fed it into waiting barges. The rivers' presence also were felt hundreds of miles away. In the 1970s grain companies began constructing inland subterminals—elevators with large storage capacities equipped to receive grain from numerous smaller, country elevators. A subterminal could load fifty, one hundred-ton railroad grain cars with corn or soybeans. Entire trainloads moved from the subterminals to one of the larger barge-loading facilities along the Mississippi River equipped to dump grain cars. When the export business was slow, subterminals provided strategic storage sites from which grain could move to various domestic destinations (especially to Southern poultry feed processors) as needed.

Missouri and southern Illinois farmers shifted heavily into soybeans. Local processors took about half their output, and the rest was shipped downriver. A Missouri farmers' association operated river terminals on the Missouri River, and the state of Indiana did likewise on the Ohio. Western Tennessee and Kentucky increased their soybean fields by 750,000 acres during the 1970s. An even larger expansion took place in the Missouri Bootheel and the St. Francis Basin of Arkansas, which already had the heaviest concentrations of soybeans in the United States. As with corn farther to the north, beans

Fig. 55. Mississippi River grain-barge terminal, Whiteside County, Illinois. After more than a century of relative unimportance, the Mississippi River again became an important route to market for Corn Belt farmers during the grain-export boom of the 1970s.

were trucked to elevators on the Mississippi and its tributaries, then poured into barges and sent to the vicinity of New Orleans.

The export market reached its peak in 1979 and declined steadily for the next six years. France, Brazil, Japan, and Argentina began to develop more aggressive trade policies beginning in 1980, when U.S. corn and soybean prices were higher than those of competing countries. The grain export boom took less time to collapse than it had taken to build, and it did so with disastrous consequences. Grain prices quickly dropped to 1960s levels, making it impossible for many farmers to continue paying back the loans they had received to expand their operations during the inflationary 1970s. The supply of credit began to shrink.

Agricultural economies in corn-dependent states were severely impacted. According to Mark Friedberger the farm crisis in Iowa unfolded in several stages.[37] Deflation of farmers' assets began in 1982 and led to a period of confrontation and the buildup of farm advocacy. The staggering amount of farm indebtedness that overwhelmed creditors eventually had to be restructured.[38] Farmers began to mobilize for their own cause in 1985, but by 1986 they were learning to accept what had happened: a 55% write-down in the value of their assets in just four years. In Friedberger's opinion farm advocacy only stalled the inevitable, and, while it gave farmers a temporary cushion, a more "abrupt departure to get on with another career" would have been the more realistic course to take. A "Save the Family Farm" movement followed the crisis of the 1980s but had little impact.[39]

Farm policy directed at the Corn Belt in recent years has been largely concerned with controlling the problems of overproduction that began to appear as the export boom collapsed. The PIK (Payment-In-Kind) Program diverted seventy-eight million acres out of crop production in 1983. PIK certificates entitled farmers to a specified amount of stored, surplus grain in lieu of the crops they would have grown themselves. It was an attempt to deal with the bulging surplus that appeared in the 1980s. The annual carryover in U.S. grain stocks exceeded one hundred million metric tons in 1981 and surpassed two hundred million metric tons by 1986.[40] In 1990 the maximum payment acreage a single farmer could have was placed under limitations. Grain dealers also have dealt in PIK certificates, since provisions of the 1985 Farm Bill (Food Security Act of 1985) included certificates as part of the government's Export Enhancement Program. Grain exporting firms are paid in PIK certificates the balance between the price they can obtain selling grain in a foreign country and the cost of that grain to them plus a fair profit.

Acreage restrictions continue to be a major aspect of federal programs to limit production. More than 35 million acres of erodible cropland were placed in the Conservation Reserve Program between 1980 and 1991; soil conservation measures were explicitly made part of the 1985 Farm Bill. The old Corn Belt fringe, including the southern Iowa/northern Missouri area of clay subsoils and extreme western Kentucky, has noticeably been withdrawn from crop production.[41] But the combination of conservation reserve and export enhancement programs has to be managed carefully. A total of 60.5 million acres of cropland were idle under one or another program in 1987, but by 1992 grain stocks were low as a result of several droughts and decreased acreage during the intervening years. Off-farm corn stocks declined 80% between 1987 and 1990. Part of the drop was due to an increase in corn exports, the result of large purchases by the Soviet Union and Japan in 1988–1989. The former Soviet Union continues to purchase U.S. grain now only under the system of export credit guarantees.[42]

Corn exports have fluctuated over the past several decades, but the total share that is exported has remained fairly stable. Disposition of the corn crop remains heavily dependent on feed grain uses; about five acres of every eight planted in corn still goes for feed. Apart from on-farm consumption by hogs and cattle, substantial quantities of corn and soybeans go to the U.S. poultry industry. The other three acres in eight produce corn, in roughly equal amounts, either for export or for domestic corn milling industries. Partly because of the stagnation in exports, but also because of its many new uses, the proportion of U.S. corn consumed in wet milling has more than doubled, from 5.6% in 1975 to 13% in 1989.[43]

Another twist in Corn Belt logistics has come from the changing global economy. Partial deregulation of railroads by the federal government in 1976 led to new pricing strategies for volume, long-distance shipments on dry-bulk commodities such as grain. Because there has been a large demand for feed grains in the Far East, and given the circuity of water routes reaching there from the mouth of the Mississippi River, railroad companies began to set low-cost rates on grain shipments from the Corn Belt to Pacific Northwest ports.[44] Rail rates on grain are sensitive to distance, and therefore much of the corn that moves to ocean ports on Puget Sound or the Columbia River originates in the western Corn Belt.

Nebraska is now the third largest corn-producing state. Its early Corn Belt, developed in the southeastern counties by the 1870s, is all but forgotten now that much of the state's crop is grown under irrigation. Nebraska's corn is sold to Great Plains feedlots and Pacific Rim countries, buyers unknown in the nineteenth century. Feedlots have to pay higher prices for corn in years of large exports to the Far East because the Platte Valley and the Nebraska/Colorado High Plains are in the front line of supply to Pacific Coast ports. Trainloads of corn and soybeans also move toward Pacific markets from western Minnesota and southeastern North Dakota. Feed-grain demand in the Far East is exerting a strong westward pull on U.S. corn production, which has helped stretch the western Corn Belt from the Texas Panhandle to the Red River Valley of North Dakota (fig. 55).

The demand for corn in Nebraska is so strong that, on the state's irrigated acreage, 75% of the cornfields are used to plant corn every year. Continuous corn accounts for less than 30% of the acreage in Illinois and Iowa, however, where about 60% of the fields have two years of corn followed by one of soybeans. Nearly all the acreage is treated with herbicides and is heavily fertilized.[45] Crop and livestock farming have separated to the point where only one Corn Belt acre in six now receives animal manure. At the turn of the century Leonidas Kerrick, the McLean County, Illinois, cattle feeder, had predicted that cash-grain farming would rob the land of its fertility. No good agriculture was disassociated from the raising and feeding of livestock, Kerrick thought. "If he robbed the land, he robbed his children."[46]

If no resources from one place are ever to be used in another, then Kerrick was right: what is taken from a piece of land must, in some way, be returned to it, from that locality. If those resources are exhaustible, then agriculture is not sustainable. But the Corn Belt's history has been one of taking more than it has been one of replacing. The northward movement of corn in the late nineteenth and early twentieth centuries was a net shift away

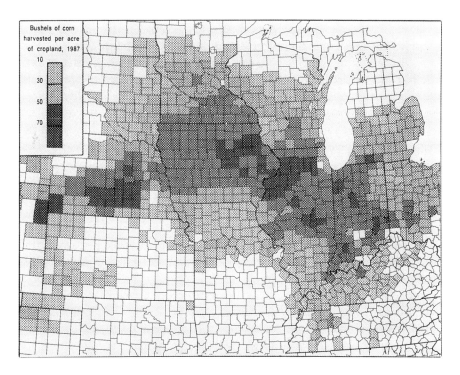

Fig. 56. Bushels of corn harvested per total cropland acres (U.S. Census of Agriculture, 1987).

from lands that no longer produced large corn crops. Farming has been sustained only because it uses rocks dug in, say, Florida to replenish soil fertility or hydrocarbons extracted in Kuwait to effect that replenishing. Corn requires 40% of the fertilizer used on farms in this country and nearly an equal percentage of the herbicide. Corn produced on these farms is consumed more or less everywhere. From whence, then, shall these lands be replenished?

The well-known patterns across the region thus continue: good land, prairies, drained land, stock feeding, corn and swine production, soybeans, cash grain farming, and tenancy—all have, or have had, maps that resemble that of corn production or chemical consumption in 1987. The correspondences, conjunctions both in time and over the years, are what define the Corn Belt as a region and simultaneously verify our sense of its significance. The map of corn production in 1987 looks like earlier years, but the numerical legend is more than double that of the 1949 map (figs. 1 and 56).

The future will see a continuation of some of these patterns, although Corn Belt history has been notably difficult to predict. Whatever the next

revolution brings, it will likely emphasize sustainable agriculture and will attempt to diverge from the current attachment to agrochemicals and fossil
fuels. According to the U.S. Department of Agriculture, sustainable agriculture has four distinguishing practices: crop rotation; reduced tillage (to leave
more crop residues on the soil surface); maintenance of soil fertility (such as
the use of manure); and the incorporation of pest control procedures that
are less harmful to the environment.[47] These are mostly old ideas, not new
ones, and they were once practiced in the Corn Belt to a far greater extent
than they are today.

Sustainable agriculture is considered by some observers to be a necessary
part of world-wide adjustments in food production systems to the resources
of given environments.[48] Others who favor the idea see it quite differently.
From a biotechnology perspective "sustainable agriculture is profitable and
environment-friendly; conserves renewable and nonrenewable resources, germ
plasm, and technology; and meets consumer and social needs." An emphasis
on biological, rather than chemical, systems would reduce the consumption
of fossil-fuel energy. Biopesticides ("microbial control agents") would replace
chemical pesticides, while genetically engineered seeds would take the place
of the less sophisticated ones now in use.[49] In the past, corn varieties selected
only for yield had added problems with insect pests. Fertilized crops like
corn were especially attractive to aphids, for example.[50] In the biotechnolog-
ical version of sustainable agriculture those crops that are grown are the ones
that fit available technology for their production and consumption.

Predictions of global warming within the next half-century have caused
numerous speculations concerning a rearrangement of environments on the
earth. Although an increase in atmospheric carbon dioxide would increase
green-plant growth, the increased surface temperatures that result from added
greenhouse gases would tend to limit such growth.[51] Since global warming
does not suggest any change in earth-sun relations, daylengths and sun an-
gles—both critical for corn varieties—would remain constant. But much hotter summer temperatures, combined with drier conditions, would require
shorter-season hybrids. The main limitation on corn production would be
longer and more frequent episodes of hot summer temperatures. Irrigation
would thus be increasingly in demand, even though it is the burning of fossil
energy necessary to operate things like irrigation pumps that has brought
about the very condition of concern.

The second hybrid-corn revolution was based on improved seed but also
required the adoption of still more intensified farming practices, especially
the use of large machinery and the application of agricultural chemicals. The
first revolution, begun in the 1920s, was a commercial success by 1936 but
ended in the 1950s when the limits of production became clear. Among other

things, farmers learned that hybrid corn performed especially well if fertilizer inputs were increased. It seems there has always been something new for farmers to learn about and then, oftentimes, adopt all at once. Soybeans gave them a valuable rotation crop to plant in fields suddenly idled when oats were no longer needed. Krug corn was the hope of the 1920s. Before that it was refrigerator cars, railroads, Poland-China hogs, and steamboats. The system got organized in the five islands of good land west of the Appalachians, where people first raised large corn crops to feed to cattle and swine.

The Corn Belt has changed and so have the five islands. Ohio's Pickaway Plains and most of the Virginia Military District still are important parts of the Corn Belt. Their fertile soils are as prized today as they were nearly two centuries ago when the Renick clan and others from the South Branch sliced back the sod on grassy bottoms and planted corn. Miami Valley agriculture is now hemmed in by urban sprawl and, in large areas, has disappeared entirely. The same is true of the Bluegrass, long known for its horse farms, now also a landscape of exurban industries and commuter villages, even though the stone fences built by slaves still can be seen.[52] Tobacco is its main crop. The Pennyroyal remains rural with a mixed agriculture that includes tobacco and Corn Belt crops; its history as a "barren" still can be sensed by a person looking across rolling uplands. Apart from a few sausage and bacon factories the Nashville Basin yields few clues that it was once the center of hog production in the United States.

Most beef cattle, East or West, are now fed corn in Western-style drylots (seldom dry in the middle-western climate), whose stench is sufficient to warn away the curious. Much of what used to be visible is gone or shielded from view. Hogs are now fed in larger-scale versions of poultry houses that they do not leave until taken to slaughter. Union stockyards have nearly disappeared. Some of the largest meat-packing plants in the world still are located in the Corn Belt, but they look no different from other two-story, metal-sided industrial buildings. Thousands of refrigerated trucks, indistinguishable from any others on the highway, are at all times shuttling between the packing plants and supermarket chains across the country.

Farm residents now account for less than 2% of the national population. There were only one-fourth as many farm residents in the middle-western states in 1991 as there had been in 1920, although in the nation as a whole the decline amounted to 85% of the 1920 total.[53] To be a "farm resident," even, no longer means what it once did. In the Middle West today, a slight majority of farm residents have their principal employment off the farm. Although nearly two-thirds of the men work primarily on the farm, nearly three-fourths of farm women have their principal employment elsewhere. Farmers have left the land in droves, but those who remained have produced even more than before. The 75% decline in middle-western farm residents between 1920 and the present was accompanied by a 1,300% increase in the region's corn crop.

The "typical" Corn Belt farm has vanished and has been replaced by a small cluster of metal buildings surrounding a suburban-type tract home. The corn and soybean farmer has no need for barns, fences, or pasture. Machine sheds and grain bins are all that is required. Small grain elevators found in towns spaced five to ten miles apart have been torn down and replaced by more widely spaced subterminals that can store ten times as much grain. The whole countryside has been cleared out, leaving only the quiet monotony of fields. The most dramatic evidence of the Corn Belt's importance is in the cities, corn- and soybean-processing centers like Decatur, Illinois, where Staley and Archer-Daniels-Midland seem to have nearly taken over the city, from downtown to the northeast side, and are still building. Steam hisses and aromas pour constantly forth from the huge factories.

For many years a stone marker at Funks Grove, Illinois, indicated the spot where, in the 1820s, Isaac and Cassandra Funk had built their first home. The marker has been moved to a nearby park because the interstate highway was built through the original site. Here, surely, is the heart of the Corn Belt—a middle-western prairie settled early by cattle feeders who moved in from the Virginia District. One does not expect a monument to so large and diversified a thing as nineteenth-century Corn Belt agriculture. In many respects there is no need for a memorial, since evidence of what was established more than a century ago still is visible on all sides. But times have changed. At the south end of Funks Grove a sign is posted designating "Funks Grove Natural Area." Commemorating the Funks? No. The area is being "managed to restore it to its original natural character," "the way it would have appeared before the settlement of white people."

Presumably those responsible do not intend to erase the memory of Isaac and Cassandra Funk by restoring their property to the way it might have looked before the Funks ever saw it. There was no "original" natural character to this place, of course. We find prairies on woodland soils in some parts of mid-America, forests on prairie soils in others. At an unknown time during the past several thousand years people in Illinois began firing the land. At Funks Grove a prairie was created and a stretch of rough moraine protected the grove of trees. The Funks needed both wood and grazing land; like so many others, they took up residence at the edge of the woods. Funk's cattle grazed meadows, opened out of woodland by fires the Indians had set for hunting bison unknown numbers of years earlier. The landscape now being restored to its "natural" state was itself largely the product of Native

American lifeways, as much the result of human effort as is the present landscape of farms.

More and more we would like to have things the way they were "before," when they were original and, hence, natural. We seek a benchmark by which we can measure how far we have gone. Such yearnings unfortunately make it more difficult to assess the Corn Belt's significance in history and also tend to cloud any view of its future. This agricultural system sprang rapidly not from nature but on a landscape already improved by the works of people. For more than 150 years it has been a region of experimentation with exotic plants and animals: maize from Mexico, soybeans from China, sorghums from Africa, pigs from China and Europe, and cattle from western Europe. At present it seems to have reached a dead end, with no new promising technologies on the horizon or really any new ideas entering the discussion except from those who reasonably urge a return to cultures less harmful to the land. Microbial control agents are no more welcomed by most people than are tanks of 2–4D. Even though we do not talk this way anymore, the Corn Belt is still the "feedbag of democracy." For the near future, at least, it seems destined to remain that way.

Notes

1. Corn Belt Geography

1. William Warntz, "An Historical Consideration of the Terms, 'Corn' and 'Corn Belt' in the United States," *Agricultural History* 31 (1957), 40–45; James R. Shortridge, *The Middle West: Its Meaning in American Culture* (Lawrence: University Press of Kansas, 1989), chap. 2. The twelve states included in the Census Bureau's Middle West (formerly, North Central states) are Ohio, Indiana, Michigan, Wisconsin, Illinois, Minnesota, Iowa, Missouri, North Dakota, South Dakota, Nebraska, and Kansas.

2. Ladd Haystead and Gilbert C. Fite, *The Agricultural Regions of the United States* (Norman: University of Oklahoma Press, 1955), chap. 7. For other definitions of the region see John H. Garland, ed., *The North American Midwest* (New York: Wiley, 1955).

3. Oliver E. Baker, "Agricultural Regions of North America, Part I—The Basis of Classification," *Economic Geography* 2 (1926), 459–93; Baker, "Part IV—The Corn Belt," *Economic Geography* 3 (1927), 447–65.

4. Oliver E. Baker, "Part III—The Middle Country Where South and North Meet," *Economic Geography* 3 (1927), 309–39; U.S. Department of Agriculture, *Generalized Types of Farming in the United States*, Agriculture Information Bulletin no. 3 (Washington, D.C., 1950).

5. Environmental factors and limits on crop production are presented in Karl H. W. Klages, *Ecological Crop Geography* (New York: Macmillan, 1942).

6. J. E. Spencer and Ronald J. Horvath, "How Does an Agricultural Region Originate?" *Annals*, Association of American Geographers, 53 (1963), 74–92; quotations from 81–82.

7. John Fraser Hart, *The Land That Feeds Us* (New York: Norton, 1991), chaps. 1, 6.

8. The most generally accepted cartographic definition of the Upland South is found in Fred B. Kniffen, "Folk Housing: Key to Diffusion," *Annals*, Association of American Geographers, 55 (1965), 549–77; see also Milton B. Newton, Jr., "Cultural Preadaptation and the Upland South," in *Man and Cultural Heritage*, ed. H. J. Walker and W. G. Haag, *Geoscience and Man*, vol. 5 (Baton Rouge: School of Geoscience, Louisiana State University, 1974), 143–54; a critique of Newton's version is found in Terry G. Jordan and Carl Parker, "Commentary: Southern Folk Housing," *Annals*, Association of American Geographers, 68 (1978), 448–50.

9. On the origins of folk buildings, see Terry G. Jordan and Matti Kaups, *The American Backwoods Frontier* (Baltimore: Johns Hopkins University Press, 1989). European cattle brought to the New World are illustrated in Marlene Felius, *Genus* Bos (Rahway, N.J.: Merck, 1985). See also Charles Wayland Towne and Edward Norris Wentworth, *Cattle and Men* (Norman: University of Oklahoma Press, 1955), and Towne and Wentworth, *Pigs from Cave to Corn Belt* (Norman: University of Oklahoma Press, 1955). Histories of corn and hogs that discuss the Upland South include Sam B. Hilliard, *Hog Meat and Hoecake* (Carbondale: Southern Illinois University Press, 1972); Howard T. Walden, *Native Inheritance: The Story of Corn in America* (New York: Harper and Row, 1966); and Nicholas P. Hardeman, *Shucks, Shocks, and Hominy Blocks: Corn as a Way of Life in Pioneer America* (Baton

Rouge: Louisiana State University Press, 1981). Genetic differences between Flints and Dents, the two major races of corn important for the Corn Belt, are described in William L. Brown and Edgar Anderson, "The Northern Flint Corns," *Annals of the Missouri Botanical Garden* 34 (1947), 1–28, and in Brown and Anderson, "The Southern Dent Corns," *Annals of the Missouri Botanical Garden* 35 (1948), 255–68. Betty Fussell, *The Story of Corn* (New York: Knopf, 1992), is by far the best contemporary account.

10. Percy Wells Bidwell and John I. Falconer, *History of Agriculture in the Northern United States, 1620–1860* (Washington, D.C.: Carnegie Institution of Washington, 1925), chap. 36; Lewis Cecil Gray, *History of Agriculture in the Southern United States to 1860* (Washington, D.C.: Carnegie Institution of Washington, 1932), vol. 1, 841–77; George F. Lemmer, "The Spread of Improved Cattle through the Eastern United States to 1850," *Agricultural History* 21 (1947), 79–93. Paul C. Henlein, *Cattle Kingdom in the Ohio Valley, 1783–1860* (Lexington: University Press of Kentucky, 1959).

11. Because both the crop and its use figure in the Corn Belt's definition, indexes based on corn acreage can be misleading. Dairy farmers in the Upper Middle West often plant large acreages in corn, but they harvest it green for silage. See, for example, the "corn-oats-hay" region of the Middle West, described by John C. Weaver, which extended from northern Wisconsin to southern Iowa; the break between dairy farming and hog feeding is invisible (Weaver, "Crop Combination Regions in the Middle West," *Geographical Review* 44 [1954], 175–200; and "Crop Combination Regions for 1919 and 1929 in the Middle West," *Geographical Review* 44 [1954], 560–72).

12. The definition of improved farmland was not totally consistent in the censuses from 1850 to 1920. In 1850 improved land was that producing "crops or in some manner [adding] to the productions of the farmer" (7th Census, 1850, *Abstract*, 47). In 1870 (9th Census, vol. 3, *Wealth and Industry*, Agriculture, p. 71), "land used for grazing, grass, tillage, or lying fallow" was specifically included. In 1900 (12th Census, vol. 5, *Agriculture*, part 1, xix) it was reported that meadows and pastures were double-counted in some areas in 1880. In 1910 and 1920 (the last year of reporting improved farmland) "woodland" was specifically included under unimproved land. Regardless of inconsistencies, however, improved land may be considered to embrace all of the land a reporting farmer regarded as cultivated, fenced, mowed, cleared, drained, or otherwise the product of labor and clearly not confined just to croplands in any of the censuses, 1850–1920. Because improved farmland was not reported in 1840, the Corn Belt for that year was defined as counties meeting the second test (corn/livestock ratio) in 1840 and the first test in 1850.

13. The practices involved are described in Carl E. Leighty, "Crop Rotation," in *Soils and Men: Yearbook of Agriculture, 1938* (Washington, D.C.: U.S. Department of Agriculture, 1938), 406–30.

14. "Yankee" is used here to describe persons born in New England and New York. See John C. Hudson, "Yankeeland in the Middle West," *Journal of Geography* 85 (1986), 195–200; and Hudson, "North American Origins of Middlewestern Frontier Populations," *Annals*, Association of American Geographers, 78 (1988), 395–413. Population patterns are discussed in Robert P. Swierenga, "The Settlement of the Old Northwest: Pluralism in a Featureless Plain," *Journal of the Early Republic* 9 (1989), 73–106.

15. Eric Foner, *Free Soil, Free Labor, Free Men: The Ideology of the Republican Party before the Civil War* (New York: Oxford University Press, 1970), 49; Allan G. Bogue, *From Prairie to Corn Belt: Farming on the Illinois and Iowa Prairies in the Nineteenth Century* (Chicago: University of Chicago Press, 1963), 20. Foner is singled out not because he originated the notion, but rather because he repeated it succinctly.

16. William Cronon, in *Nature's Metropolis: Chicago and the Great West* (New York:

Norton, 1991), 46–54, revives the idea that nineteenth-century Chicago developed as a sort of focus of concentric rings of land-use intensity resembling the location-theoretic model of von Thunen (see Johann Heinrich von Thunen, *Von Thunen's Isolated State* [1842], ed. Peter Hall, trans. Carla M. Wartenberg [New York: Pergamon Press, 1966]). Any resemblance to Thunen rings is more likely due to a northward-sloping temperature gradient and a westward-sloping precipitation gradient than it is to the economic-rent mechanism Thunen postulated. Harold H. McCarty and James B. Lindberg, *A Preface to Economic Geography* (Englewood Cliffs, N.J.: Prentice-Hall, 1965), fig. 3.3., show a Thunen land-use model superimposed on environmental gradients.

17. Richard K. Vedder and Lowell E. Galloway, "Migration in the Old Northwest," in David C. Klingaman and Richard K. Vedder, eds., *Essays in Nineteenth Century Economic History: The Old Northwest* (Athens: Ohio University Press, 1975), 159–76; 173.

18. John Filson, *The Discovery and Settlement of Kentucke* [1784] (Ann Arbor: University Microfilms, 1966), 24. By the early 1850s corn yields had declined to 50 bushels to the acre in the Inner Bluegrass around Lexington; *Report of the U.S. Commissioner of Patents, 1850*, part 2, Agriculture (Washington, D.C.: 1851), 22.

19. The usual role assigned to people is one of land-clearing, but usually with little explanation. Bogue, *Prairie to Corn Belt*, 5, states that "for generations pioneers had slashed out their farms from rolling woodlands." Gilbert Fite refers to the "grueling work" of clearing land (Fite, "The Pioneer Farmer," in Vivian Wiser, ed., *Two Centuries of American Agriculture* [Washington, D.C.: Agricultural History Society, 1976], 275–89). While the costs of establishing a successful farm were substantial, it is difficult to imagine many pioneers trying to make a farm out of the timber. On farm-making costs see Clarence Danhof, *Change in Agriculture: The Northern United States, 1820–1870* (Cambridge: Harvard University Press, 1969).

20. Alfred W. Crosby, *Ecological Imperialism: The Biological Expansion of Europe, 900–1900* (Cambridge: Cambridge University Press, 1986), 147. Cf. Gary S. Dunbar, "Henry Clay on Kentucky Bluegrass, 1838," *Agricultural History* 51 (1977), 520–23.

2. Making the Land

1. D. M. Mickelson et al., "The Late Wisconsin Glacial Record of the Laurentide Ice Sheet in the United States," in H. E. Wright, Jr., ed., *Late-Quaternary Environments of the United States*, vol. 1, *The Late Pleistocene*, Stephen C. Porter, ed. (Minneapolis: University of Minnesota Press, 1983), 3–37.

2. Robert V. Ruhe, "Depositional Environment of Late Wisconsin Loess in the Midcontinent United States," in Wright, *The Late Pleistocene*, 130–37; and Ruhe, *Quaternary Landscapes in Iowa* (Ames: Iowa State University Press, 1969), 28–53.

3. John T. Curtis, *The Vegetation of Wisconsin* (Madison: University of Wisconsin Press, 1959), 254.

4. Hazel R. Delcourt and Paul A. Delcourt, *Quaternary Ecology* (London: Chapman and Hall, 1991), chap. 3.

5. Russell W. Graham, "Plant-Animal Interactions and Pleistocene Extinctions," in D. K. Elliott, ed., *Dynamics of Extinction* (New York: Wiley, 1986) 131–54.

6. Paul S. Martin, "Prehistoric Overkill: The Global Model," Martin and Richard G. Klein, eds., *Quaternary Extinctions* (Tucson: University of Arizona Press, 1984) 354–97; tables 17.2–17.3.

7. Holmes A. Semken, Jr., "Holocene Mammalian Biogeography and Climatic

Change in the Eastern and Central United States," in *Late-Quaternary Environments*, vol. 2, *The Holocene*, H. E. Wright, Jr., ed. (Minneapolis: University of Minnesota Press, 1984), 182–207; fig. 2.

8. The issue of when people first moved into North America is far from being settled. Reviews of the subject include William N. Irving, "Context and Chronology of Early Man in the Americas," *Annual Review of Anthropology* 14 (1985), 529–55; David J. Meltzer, "Why Don't We Know When the First People Came to North America?" *American Antiquity* 54 (1989), 471–90; and Frederick Hadleigh West, "The Antiquity of Man in America," in Wright, vol. 1, *The Late Pleistocene*, 364–82.

9. J. Stoddard Johnston, *First Explorations of Kentucky: Journals of Thomas Walker, 1750, and Christopher Gist, 1751*, Filson Club Publications, vol. 13 (Louisville: John P. Morton, 1898), Gist's diary, March 8, 1751.

10. For a review of the debate, see Simon J. M. Davis, *The Archaeology of Animals* (New Haven: Yale University Press, 1987). J. E. Mosimann and P. S. Martin, "Stimulating Overkill by Paleoindians," *American Scientist* 63 (1975), 304–13, offers a plausible simulation of megafaunal extinction that begins with a small band of hunters moving south from the vicinity of Edmonton ca. 11,500 B.P.

11. West, "Antiquity of Man"; Ernest L. Lundelius et al., "Terrestrial Vertebrate Faunas," in Wright, vol. 1, *The Late Pleistocene*, 311–53.

12. My interpretation of vegetation change follows T. Webb, III, et al., "Holocene Changes in the Vegetation of the Midwest," in Wright, vol. 2, *The Holocene*, 152–65, who presents isopolls based on reconstructions at 49 sites ranging from Ohio to Minnesota. I have inferred that the prairie was "present" at a site when 10% of the pollen grains were prairie forbs or grasses.

13. F. W. Albertson and J. E. Weaver, "Injury and Death or Recovery of Trees in Prairie Climate," *Ecological Monographs* 15 (1945), 395–433; and J. E. Weaver, *North American Prairie* (Lincoln, Nebr.: Johnsen, 1954), chap. 12.

14. A strictly climatic interpretation of grasslands is presented in Jack R. Harlan, *Theory and Dynamics of Grassland Agriculture* (Princeton: Van Nostrand, 1956). Most Quaternary researchers regard expanding grasslands as the product of climatic change to warmer and drier conditions, which, in turn, are inferred from the composition of pollen rains and vertebrate faunal remains, from changes in stream channels, and sometimes from variations between archaeological horizons. As an example, see Wayne M. Wendland et al., "Evaluation of Climatic Changes on the North American Great Plains Determined from Faunal Evidence," in Russell W. Graham et al., eds., *Late Quaternary Mammalian Biogeography and Environments of the Great Plains and Prairies*, Illinois State Museum, Scientific Papers, vol. 22 (1987), 460–72; James E. King, "Late Quaternary Vegetation History of Illinois," *Ecological Monographs* 51 (1981), 43–62; and H. E. Wright, Jr., "History of the Prairie Peninsula," in Robert E. Bergstrom, ed., *The Quaternary of Illinois*, Special Publication no. 14 (Urbana: University of Illinois, College of Agriculture, 1968), 78–88.

15. Reid A. Bryson and F. Kenneth Hare, "The Climates of North America," in Bryson and Hare, eds., *Climates of North America*, World Survey of Climatology, vol. 11 (Amsterdam: Elsevier, 1974), 1–47.

16. James C. Knox, "Responses of River Systems to Holocene Climates," in Wright, vol. 2, *The Holocene*, 26–41; Webb et al., fig. 3.

17. John R. Borchert, "Some Regional Differences in the Atmospheric Circulation over North America," *Annals*, Association of American Geographers, 38 (1948), 53–54; the major statement is Borchert, "The Climate of the North American Grassland," *Annals*, Association of American Geographers, 40 (1950), 1–39.

18. Webb et al., "Vegetation Changes," 163; Curtis, *Vegetation of Wisconsin*, 136; Weaver, *North American Prairie*, 191–92.

19. Stephen J. Pyne, *Fire in America* (Princeton: Princeton University Press, 1982), chap. 2, offers many historical examples. Prehistoric fires have been dated with radiocarbon techniques. Near Cherokee, in northwestern Iowa, the early Holocene forest was destroyed by fires roughly 8000 B.P.; see R. G. Baker and K. L. Van Sandt, "Holocene Vegetational Reconstruction in Northwestern Iowa," in Duane C. Anderson and Holmes A. Semken, Jr., eds., *The Cherokee Excavations* (New York: Academic, 1980), 123–38.

20. Carl O. Sauer, "A Geographic Sketch of Early Man in America," *Geographical Review* 34 (1944), 529–73. The many possible causes of prairie vegetation suggested by early travelers as well as by some botanists and geographers are reviewed in Douglas R. Mc-Manis, *The Initial Evaluation and Utilization of the Illinois Prairies*, Research Paper no. 94 (Chicago: University of Chicago, Department of Geography, 1964), 13–16. In *Fire in America*, Stephen Pyne states (p. 81), "the evidence for aboriginal burning . . . is so conclusive . . . that it seems fantastic that a debate about whether Indians used broadcast fire or not should ever have taken place." Also see Omer C. Stewart, "Fire as the First Great Force Employed by Man of the Earth," in William L. Thomas, ed., *Man's Role in Changing the Face of the Earth*, vol. 1 (Chicago: University of Chicago Press, 1956), 115–33; and William M. Denevan, "The Pristine Myth: The Landscape of the Americas in 1492," *Annals*, Association of American Geographers, 82 (1992), 369–85.

21. W. T. Hornaday, "The Extermination of the American Bison, with a Sketch of Its Life History," *Annual Report*, U.S. National Museum (Washington, D.C., 1880), 367–548. Representative studies of bison and the settlement process in various areas include Tom D. Dillehay, "Late Quaternary Bison Population Changes on the Southern Plains," *Plains Anthropologist* 19 (1974), 180–96; John A. Jakle, "The American Bison and the Human Occupance of the Ohio Valley," *Proceedings of the American Philosophical Society* 112 (1968), 299–305; and Erhard Rostlund, "The Geographic Range of the Historic Bison in the Southeast," *Annals*, Association of American Geographers, 50 (1960), 395–407.

22. Carl O. Sauer, *Seventeenth Century North America* (Berkeley: Turtle Island, 1980), 226, describes changes in the Texas Coastal Plain between 1535, when survivors of the Narvaez expedition lived there, and 1687, when LaSalle's party saw it.

23. Webb et al., "Vegetation Changes," fig. 4.

24. Reuben Gold Thwaites, ed., *The Jesuit Relations and Allied Documents*, vol. 59 (Cleveland: Burrows Bros., 1900), 147. The French called the wild bison "Illinois cattle" and learned to anticipate seeing them in quantity by the time they were as far down the Ohio as the mouth of the Kanawha (*Jesuit Relations*, vol. 69, 177).

25. John Filson, *The Discovery and Settlement of Kentucke* [1784] (Ann Arbor: University Microfilms, 1966), 23. Some Ohio cattlemen thought bluegrass was native to their state, but not to Kentucky; see William Renick, *Memoirs, Correspondence, and Reminiscences* (Circleville, Ohio: privately printed, 1880), 33–56. Some Kentuckians thought it had been introduced there from Lancaster County, Pennsylvania, in 1752; Lewis Cecil Gray, *History of Agriculture in the Southern United States, to 1860* (Washington, D.C., Carnegie Institution of Washington, 1932), vol. 1, 861. Bluegrass appeared spontaneously and under cultivation in southeastern Pennsylvania before 1700 (James T. Lemon, *The Best Poor Man's Country* [Baltimore: Johns Hopkins University Press, 1972], 159). Alfred Crosby, *Ecological Imperialism: The Biological Expansion of Europe, 900–1900* (Cambridge: Cambridge University Press, 1986), 157, states that Americans "arrogantly" named it Kentucky Bluegrass.

26. Brent Altsheler, "The Long Hunters and James Knox, Their Leader," *Filson Club History Quarterly* 5 (1931), 173.

27. Thomas Perkins Abernethy, *From Frontier to Plantation in Tennessee* (Chapel Hill: University of North Carolina Press, 1932), 29. The Bluegrass and the Nashville Basin share a similar mix of hardwood species not found elsewhere, which cannot be explained in purely natural terms (E. Lucy Braun, *Deciduous Forests of Eastern North America* [Philadelphia: Blaikston, 1950], 125–27). The Nashville Basin's cedar glades are natural, part of a xerophytic community on sandstones (Braun, 131).

28. "Filson's First Map of Kentucky," *Kentucke*, foldout map; Altsheler, "Long Hunters," 171–74. On "barrens" in Virginia, see Robert Beverley, *The History of Virginia*, 2d ed. [1772] (Richmond: J. W. Randolph, 1855), 104.

29. Nathaniel S. Shaler, *Kentucky: A Pioneer Commonwealth* (Boston: Houghton, Mifflin, 1888), 27–29; Carl O. Sauer, *Geography of the Pennyroyal*, series 6, vol. 25 (Lexington: Kentucky Geological Survey, 1927).

30. Filson, *Kentucke*, 21.

31. Gilbert Imlay, *A Topographical Description of the Western Territory of North America* (London: J. Debrett, 1797), 29–32; Timothy Flint, *A Condensed Geography and History of the Western States*, vol. 2 (Cincinnati: William M. Farnsworth, 1828), 174; Willard Rouse Jillson, ed., *Filson's Kentucke* (Louisville: John P. Morton, 1929), 122.

32. Jerry E. Clark, *The Shawnee* (Lexington: University Press of Kentucky, 1977), 12; Johnston, *Explorations*, 61; William M. Darlington, *Christopher Gist's Journals* (Pittsburgh: J. R. Weldin, 1893), 131. On Shawnee settlement strategies, see George E. Hyde, *Indians of the Woodlands* (Norman: University of Oklahoma Press, 1962); 62, 67, 144–45.

33. Imlay, *Topographical Description*, 32; Darlington, *Gist's Journals*, 132. Imlay actually saw what he described in the Bluegrass. While there, according to Lawrence Elliott's flattering biography of Daniel Boone, it was Imlay who introduced John Filson to Boone. Imlay also "persuaded the ever-trustful Boone" to sell him 10,000 acres on his note alone. Later, Imlay confessed he could not make good (Elliott, *The Long Hunter* [New York: Readers Digest Press, 1976], 171). For a different view of Filson and Boone see Richard White, *The Middle Ground* (Cambridge: Cambridge University Press, 1991), 421–22.

34. Darlington, *Gist's Journals*, 117, 121; Johnston, *Explorations*, 122–23.

3. Finding the Land

1. *Secret Journals of the Acts and Proceedings of Congress* (Boston: Thomas B. Wait, 1821), 169–75.

2. J. Stoddard Johnston, *First Explorations of Kentucky: Journals of Dr. Thomas Walker, 1750, and Christopher Gist, 1751*, Filson Club Publications, vol. 13 (Louisville: John P. Morton, 1898), 49; Thomas Perkins Abernethy, *Western Lands and the American Revolution* (New York: Appleton-Century, 1937), 1.

3. Abernethy, *Western Lands*, chap. 1; Kenneth P. Bailey, *The Ohio Company of Virginia and the Westward Movement, 1748–1792* (Glendale, Calif.: Clark, 1939), chaps. 1–2; William H. Gaines, Jr., *Virginia History in Documents, 1621–1788* (Richmond: Virginia State Library, 1974).

4. Bailey, *Ohio Company*, 58–67.

5. Johnston, *Explorations*, 51.

6. Johnston, *Explorations*, 65.

7. Bailey, *Ohio Company*, 85–100; quotation from Johnston, *Explorations*, 102. On Gist's life, see Kenneth P. Bailey, *Christopher Gist: Colonial Frontiersman, Explorer, and Indian Agent* (Hamden, Conn.: Archon, 1976).

8. The Treaty of Aix-la-Chapelle (1748) left the French-British boundary in North America indefinite; for the Celoron expedition, see Reuben Gold Thwaites, ed., *The Jesuit Relations and Allied Documents: Travels and Explorations of the Jesuit Missionaries in New France, 1610–1791*, vol. 69 (Cleveland: Burrows Bros., 1940), 151–99.

9. Robert Beverley, *The History of Virginia*, 2d ed. [1772] (Richmond: J. W. Randolph, 1855), 45.

10. William M. Darlington, *Christopher Gist's Journals* (Pittsburgh: J. R. Weldin, 1893), 110. Gist's route is shown accurately in John A. Jakle, *Images of the Ohio Valley: A Historical Geography of Travel, 1740 to 1860* (New York: Oxford University Press, 1977), 167.

11. Johnston, *Explorations*, 122; William Albert Galloway, *Old Chillicothe* (Xenia, Oh.: Buckeye Press, 1934), 23, 39.

12. Johnston, *Explorations*, 124.

13. Richard P. Goldthwait et al., "Pleistocene Deposits of the Erie Lobe," in H. E. Wright, Jr., and David G. Frey, eds., *The Quaternary of the United States* (Princeton: Princeton University Press, 1965), 85–98.

14. Johnston, *Explorations*, 132–33; Thwaites, *Jesuit Relations*, vol. 69, 189.

15. George Rogers Clark to Jonathan Clark, 22 November 1772, *George Rogers Clark Papers, 1771–1781*, James Alton James, ed., Illinois Historical Society *Collections* 8 (1912), 1–2.

16. Bailey, *Ohio Company*, 96–97; *Executive Journals of the Council of Colonial Virginia*, vol. 6, 1754–1755 (Richmond: Virginia State Library, 1966), miscellaneous papers, 699.

17. Beverley W. Bond, Jr., *Foundations of Ohio*, vol. 1 (Columbus: Ohio State Archaeological and Historical Society, 1941), 186–88; *Secret Journals of Congress*, 176; 200,000 acres of the Allegheny Plateau, west of the proclamation line, was reserved for men who enlisted in the colonial militia; see Samuel M. Wilson, "George Washington's Contacts with Kentucky," *Filson Club Quarterly* 6 (1932), 215–60.

18. John Filson, *The Discovery and Settlement of Kentucke* [1784] (Ann Arbor: University Microfilms, 1966), 78. Destroying corn crops belonging to a Shawnee village was a strategy Clark had used before (*Clark Papers*, 452). There was a good network of trails linking the villages, which made it easier for Clark and his raiders to make a circuit quickly; see Frank N. Wilcox, *Ohio Indian Trails* (Cleveland: Gates, 1933).

19. Filson, *Kentucke*, 63–65.

20. William Waller Hening, ed., *Virginia Statutes* (Richmond: J. & C. Cochran, 1821), vol. 10, 8 and 564–66; *Journals of the Continental Congress*, vol. 17 (Washington, D.C.: Library of Congress, 1928), 806–808.

21. Thomas P. Abernethy, *From Frontier to Plantation in Tennessee* (Chapel Hill: University of North Carolina Press, 1932), 24; Lawrence Elliott, *The Longhunter* (New York: Readers Digest Press, 1976), 79–81, describes the Treaty of Sycamore Shoals; Henderson's Transylvania is placed in the larger context of Appalachian settlement in Donald W. Meinig, *The Shaping of America*, vol. 1, *Atlantic America, 1492–1800* (New Haven: Yale University Press, 1986), 288–95.

22. Hening, *Virginia Statutes*, vol. 9, 561–63.

23. *Continental Congress*, vol. 34, 99, 280–81. James Madison insisted that Virginia's sovereignty was absolute within the whole of its chartered bounds; Lance Banning, "James Madison and the Nationalists, 1780–1783," *William and Mary Quarterly* 40 (1983), 227–55.

24. *Continental Congress*, vol. 34, 476.

25. *Continental Congress*, vol. 26, 118–19; this is the report in which Jefferson suggested names for the states that would be created in the Northwest Territories: Sylvania, Michigania, Cherronesus, Assenisipia, Metropotamia, Pelisipia, Washington, Saratoga, Polypotamia.

26. *Continental Congress*, vol. 33, 513.

27. Beverley W. Bond, Jr., *The Correspondence of John Cleves Symmes* (New York: Macmillan, 1926), 3–8.

28. Bond, *Symmes*, 282.

29. Bond, *Symmes*, 8.

4. Zea Mays

1. Carl O. Sauer, *Sixteenth Century North America* (Berkeley: University of California Press, 1971), chap. 9.

2. *Final Report of the United States de Soto Expedition Commission*, 76th Congress, House of Representatives Document 71 (Washington, D.C., 1939); see map in Sauer, *Sixteenth Century*, 164.

3. Marquette thought these Indians raised corn because rival tribes kept them from hunting buffalo; Reuben Gold Thwaites, ed., *The Jesuit Relations and Allied Documents*, vol. 59 (Cleveland: Burrows Bros., 1940), 157.

4. David H. Dye and Cheryl Anne Cox, *Towns and Temples along the Mississippi* (Tuscaloosa: University of Alabama Press, 1990), details the complex societies found in this area. Occupancy of the Ohio Valley is discussed in George E. Hyde, *Indians of the Woodlands, from Prehistoric Times to 1725* (Norman: University of Oklahoma Press, 1962), 144–53.

5. The literature is examined critically by three authors in Richard I. Ford, ed., *Prehistoric Food Production in North America*, Anthropological Papers 75 (Ann Arbor: University of Michigan Museum of Anthropology, 1985); Charles B. Heiser, Jr., "Some Botanical Considerations of the Early Domesticated Plants North of Mexico," 57–72; Frances B. King, "Early Cultivated Cucurbits in Eastern North America," 73–97; and Patty Jo Watson, "Horticulture in the Midwest and Midsouth," 99–147.

6. C. Wesley Cowan, "Understanding the Evolution of Plant Husbandry in Eastern North America: Lessons from Botany, Ethnography and Archaeology," in Ford, ed., *Prehistoric Food Production*, 205–43.

7. Patty Jo Watson, "Prehistoric Horticulturists," in Watson, ed., *Archaeology of the Mammoth Cave Area* (New York: Academic, 1974), chap. 31. In the Eastern Woodlands cultivation began 3,000 years ago with sunflower (from the west) and sumpweed, which moved up the Mississippi to the Wabash (Richard A. Yarnell, "Native Plant Husbandry North of Mexico," in Charles A. Reed, ed., *Origins of Agriculture* [The Hague: Mouton, 1977] 861–75). Also see Gary D. Crites and R. Dale Terry, "Nutritive Value of Maygrass, *Phalaris caroliniana*," *Economic Botany* 38 (1984), 114–20; and Stuart Struever and Kent D. Vickery, "The Beginnings of Cultivation in the Midwest-Riverine Area of the United States," *American Anthropologist* 75 (1973), 1197–1220.

8. David L. Asch and Nancy B. Asch, "Prehistoric Plant Cultivation in West-Central Illinois," in Ford, ed., *Prehistoric Food Production*, 149–203. The archaeological sites studied are mostly along the lower Illinois River on floodplains or fringing bluffs.

9. Robert Beverley, *The History of Virginia* [1772] (Richmond: J. W. Randolph, 1855), 105, reported an "incredible plenty and variety" of wild grapes there. Numerous European impressions of grapes and other fruit are catalogued in Carl O. Sauer, *Seventeenth Century North America* (Berkeley: Turtle Island, 1980); also see Cowan, "Plant Husbandry," 218–20; Sauer, *Sixteenth Century North America*, 183; and Harry J. Carman, ed., *American Husbandry* (New York: Columbia University Press, 1939), 212–13. Grape vines struggled for light in a dense forest but grew rapidly at the margins of forest openings, an example of

"edge effect" as discussed in Eugene P. Odum, *Fundamentals of Ecology* (Philadelphia: Saunders, 1959), 278.

10. Richard I. Ford, "Patterns of Prehistoric Food Production in North America," in Ford, ed., *Prehistoric Food Production*, 353.

11. George W. Beadle, "The Origin of *Zea mays*," in Reed, ed., *Origins of Agriculture*, 615; also see Beadle, "The Ancestry of Corn," *Scientific American* 242 (1980), 112–19.

12. Beadle, "Origin of *Zea mays*." The human-selection theory is explained in Walton C. Galinat, "Domestication and Diffusion of Maize," in Ford, ed., *Prehistoric Food Production*, 245–78; Galinat, "The Origin of Maize," *Annual Review of Genetics* 5 (1971), 31–37; and Galinat, "The Origin of Corn," in G. F. Sprague and J. W. Dudley, eds., *Corn and Corn Improvement*, 3d ed. (Madison: American Society of Agronomy, 1988), 2–32. Others hold that maize originated naturally, as a mutation in teosinte (Hugh Iltis, "From Teosinte to Maize: The Catastrophic Sexual Transmutation," *Science* 222 [1983], 886–94; and Stephen J. Gould, "A Short Way to Corn," *Natural History* 93 [1984], 12–20). Older theories postulated that maize had an unknown, perhaps extinct, progenitor (L. F. Randolph, "II. Cytogenetic Aspects of the Origin and Evolutionary History of Corn," in George F. Sprague, ed., *Corn and Corn Improvement* [New York: Academic, 1955], 17). A respected authority, Paul C. Mangelsdorf, in *Corn* (Cambridge: Harvard University Press, 1974), suggests a hypothetical wild ancestor, as he did in his earlier work. The herbalists of postcontact Europe thought there were two types of corn, one that had come from the New World, the other from Asia; see John J. Finan, *Maize in the Great Herbals* (Waltham, Mass.: Chronica Botanica, 1950). Betty Fussell's *The Story of Corn* (New York: Knopf, 1992), 59–96, is a delightful and quite evenhanded review of the controversy over maize origins that is based in part on her interviews with Galinat and Iltis.

13. Galinat, "Domestication and Diffusion of Maize," 264–70.

14. Dee Ann Story, "Adaptive Strategies of Archaic Cultures of the West Gulf Coastal Plain," in Ford, ed., *Prehistoric Food Production*, 19–56. A map in Paul Weatherwax, *Indian Corn in Old America* (New York: Macmillan, 1954), 52, portrays a narrow corridor across the Coastal Plain as the route of maize diffusion from Mexico.

15. Bruce Trigger, *The Huron* (New York: Holt, Rinehart and Winston, 1969), 27–28; Sauer, *Seventeenth Century North America*, 103–106.

16. Edgar Anderson and William L. Brown, "Origin of Corn Belt Maize and Its Genetic Significance," John W. Gowen, ed., *Heterosis* (Ames: Iowa State University Press, 1952), 137–38.

17. Beverley, *Virginia*, 114–15; and Joseph Ewan and Nesta Ewan, eds., *John Banister and His History of Virginia, 1678–1692* (Urbana: University of Illinois Press, 1970), 40. The discussion in Lyman Carrier, *The Beginnings of Agriculture in America* (New York: McGraw-Hill, 1923), 47, is based on Beverley.

18. Mangelsdorf, *Corn*, 191; Major M. Goodman and William L. Brown, "Races of Corn," in Sprague and Dudley, eds., *Corn and Corn Improvement*, 3d ed., 65; William L. Brown, "Sources of Germ Plasm for Hybrid Corn," Eighth Annual Hybrid Corn Industry-Research Conference *Proceedings*, 1953, 12; and George F. Carter, "The Distribution of Races of Maize among the Indians of the Mississippi Valley," *Transactions*, New York Academy of Sciences, series 2, 9 (1947), 268–69.

19. Sauer, *Sixteenth Century North America*, 221–24.

20. Edgar Anderson and William L. Brown, "Corn Belt Maize," 176; Anderson and Brown, "The History of the Common Maize Varieties of the United States Corn Belt," *Agricultural History* 26 (1952), 2–7.

21. Beverley, *Virginia*, 115.

22. Anderson and Brown, "Common Maize Varieties," 5; Simon Baatz, *Venerate the Plough* (Philadelphia: Philadelphia Society for Promoting Agriculture, 1985).
23. *Annual Report of the Commissioner of Patents*, 1848 (Washington, D.C., 1849), 652–53.
24. *Report of the Commissioner of Patents*, 1849; 18–20.
25. Anderson and Brown, "Common Maize Varieties," 126.
26. Henry A. Wallace and William L. Brown, *Corn and Its Early Fathers* (Chicago: Lakeside, 1956), 80–81.
27. Robert D. Mitchell, "The Shenandoah Valley Frontier," *Annals*, Association of American Geographers, 62 (1972), fig. 1; and John H. Wheeler, *Historical Sketches of North Carolina, 1584 to 1851* (Philadelphia: Lippincott, Grambo, 1851).
28. Gershom Flagg of Vermont reported corn growing 10 to 15 feet high in Champaign County, Ohio, in 1817. Hogs, when turned loose into the corn, never tore down more than they wished to eat (G. Flagg to Artemas Flagg, June 1, 1817, in Barbara Lawrence and Nedra Branz, eds., *The Flagg Correspondence: Selected Letters, 1816–1854* [Carbondale: Southern Illinois University Press, 1986], 7).
29. The first arrivals at Ft. Boonesboro in 1775 planted corn even as they built fortifications (Malcolm Rohrbaugh, *The Trans-Appalachian Frontier* [New York: Oxford University Press, 1978], 33–34). George Rogers Clark planted corn immediately when he settled at the mouth of the Kanawha in 1772 (and in his first letter back to Spotsylvania asked his brother, who was to come west, to bring a quantity of bluegrass seed, another planting habit of Virginians) (James Alton James, ed., *George Rogers Clark Papers, 1771–1781*, Illinois Historical Society *Collections* 8 [1912], 2).
30. For example, Joliet's party saw bottomland prairies as many as twenty leagues long and three leagues wide (Sauer, *Seventeenth Century North America*, 141–42). The expedition up the Mississippi beyond St. Louis under Maj. Stephen H. Long in 1819 reported "the prairies upon this river become more numerous and extensive as we proceed upward" (quoted in William J. Peterson, *Steamboating on the Upper Mississippi* [Iowa City: State Historical Society of Iowa, 1968], 87–88).
31. The notion that prairies were avoided is assumed in Dorothy Anne Dondore, *The Prairie and the Making of Middle America* (Cedar Rapids: Torch Press, 1926), 163–65. Good critiques of the literature on prairie avoidance are found in Robert E. Warren and Michael J. O'Brien, "A Model of Frontier Settlement," in O'Brien et al., *Grassland, Forest, and Historical Settlement* (Lincoln: University of Nebraska Press, 1984), 45–49; and Terry G. Jordan, "Between the Forest and the Prairie," *Agricultural History* 38 (1964), 205–16.
32. S. P. Hildreth, *Pioneer History* (Cincinnati: H. W. Derby, 1848), 420–21.
33. Paul C. Henlein, "Journal of F. and W. Renick on an Exploring Tour to the Mississippi and Missouri Rivers in the Year 1819," *Agricultural History* 30 (1956), 174–86. Newspaper item is quoted in *Report of the Commissioner of Patents*, 1845; 181. On Felix Renick's career see Charles S. Plumb, "Felix Renick, Pioneer," *Ohio Archaeological and Historical Quarterly* 33 (1924), 16–18.
34. *Ohio State Board of Agriculture, Annual Report*, 1846 (Columbus, 1847), 59; Henry Howe, *Historical Collections of Ohio* (Cincinnati: Derby, Bradley, 1847), 401–403.

5. The Feedlot

1. James H. Madison, ed., *Heartland* (Bloomington: Indiana University Press, 1988); John Fraser Hart, *The Land That Feeds Us* (New York: Norton, 1991), chap. 8, "Heart of the Heartland."

2. Richard Lyle Power, *Planting Corn Belt Culture: The Impress of the Upland Southerner and Yankee in the Old Northwest*, Indiana Historical Society Publications, no. 25 (Indianapolis: Indiana Historical Society, 1953), 132.

3. Avery Craven, "The Advance of Civilization into the Middle West in the Period of Settlement," in Dixon Ryan Fox, ed., *Sources of Culture in the Middle West* (New York: Appleton-Century, 1934), 44–45.

4. A. Banning Norton, *History of Knox County, Ohio* (Columbus: Richard Nevins, 1862), 10; *Chicago Weekly American*, quoted in William Vipond Pooley, *The Settlement of Illinois from 1830 to 1850*, Bulletin of the University of Wisconsin, no. 1 (Madison: University of Wisconsin Press, 1908).

5. James T. Lemon, *The Best Poor Man's Country* (Baltimore: Johns Hopkins University Press, 1972), 180. Lemon gives a fuller discussion of farming mentality in "Early Americans and Their Social Environment," *Journal of Historical Geography* 6 (1980), 115–31. The transition to a market economy probably had taken place by the early 19th century; see Alan Kulikoff, "The Transition to Capitalism in Rural America," *William and Mary Quarterly* 46 (1989), 120–44; and also James Henretta, "Families and Farms: Mentalité in Pre-Industrial America," *William and Mary Quarterly* 35 (1978), 3–32. Nonetheless, Jeremy Atack and Fred Bateman, *To Their Own Soil: Agriculture in the Antebellum North* (Ames: Iowa State University Press, 1987), 273, described the antebellum middle-western farmer as a satisficer who straddled "the fence between agriculture as a way of life and as a business enterprise like all others." Studies of what could be called a Corn Belt mentalité appear in John Fraser Hart, "The Middle West," *Annals*, Association of American Geographers, 62 (1972), 271–73, and Carl C. Taylor, "The Corn Belt," in Taylor et al., *Rural Life in the United States* (New York: Knopf, 1955), 360–82.

6. David Meade Massie, *Nathaniel Massie* (Cincinnati: Robert Clarke, 1896); Beverley W. Bond, Jr., *Civilization of the Old Northwest: A Study of Political, Social, and Economic Development, 1788–1812* (New York: Macmillan, 1934); Andrew R. L. Cayton, *The Frontier Republic* (Kent: Kent State University Press, 1986); Jeffrey P. Brown, "Chillicothe's Elite: Leadership in a Frontier Community," *Ohio History* 96 (1987), 140–56.

7. Cayton, *Frontier Republic*, 52.

8. Duncan McArthur does not qualify as an Upland Southerner by birth; he was born in Duchess County, New York, in 1776 and moved to Pennsylvania as a child; L. W. Renick, compiler, *Che-le-co-the* (New York: Knickerbocker Press, 1896), 65–66. Tiffin came from England.

9. Frank Theodore Cole, "Thomas Worthington," *Old Northwest Genealogical Quarterly* 5 (1902), 35; Alfred Byron Sears, *Thomas Worthington* (Columbus: Ohio State University Press, 1958), 19. On the slavery issue see David Brion Davis, "The Significance of Excluding Slavery from the Old Northwest in 1787," *Indiana Magazine of History* 84 (1988), 75–89.

10. Ross County Northwest Territory Committee, *Chillicothe and Ross County* (Chillicothe: Works Progress Administration, 1938), 38, contrasts the Scioto Valley with northern Ohio: at Chillicothe "big stone houses built upon the hillside still look down to the town."

11. William Thomas Hutchinson, "The Bounty Lands of the American Revolution in Ohio," Ph.D. dissertation, University of Chicago [1927] (Arno Press Facsimile Edition, 1979), 185; Henry Howe, *Historical Collections of Ohio* (Cincinnati: Derby, Bradley, 1847), 401; *History of Ross and Highland Counties, Ohio* (Cleveland: W. W. Williams, 1880), 171. A breakdown into smaller farms occurred later, however; see *Atlas of Ross County Ohio* (Columbus: H. T. Gould, 1875). The landscape resulting from the Virginia survey system versus that of the Northwest Ordinance is vividly illustrated in Sam B. Hilliard, "A Robust New Nation, 1783–1820," in Robert D. Mitchell and Paul A. Groves, eds., *North America: The*

Historical Geography of a Changing Continent (Totowa, N.J.: Rowman & Littlefield, 1987), fig. 7.4.

12. Ohio State Board of Agriculture, *Third Annual Report*, 1848 (Columbus, 1849), 49.

13. Ohio Agriculture, *First Annual Report*, 1846, 26–60.

14. William Renick, *Memoirs, Correspondence, and Reminiscences* (Circleville, Ohio: privately printed, 1880), 11. This portion of Renick's book is a slight revision of his earlier "On the Cattle Trade of the Ohio Valley," Ohio Agriculture, *Third Annual Report*, 1848, 162–64.

15. Renick, *Memoirs*, 33.

16. Randolph Chandler Downes, *Frontier Ohio, 1788–1803* (Columbus: Ohio State Archaeological and Historical Society, 1935), 101–105; Renick, *Memoirs*, 12.

17. John Woods, *Two Years' Residence in the Settlement on the English Prairie* (London: Longman, Hurst, Rees, Orme, and Brown, 1822), 45–47.

18. Ohio Agriculture, *First Annual Report*, 1846.

19. Lewis Cecil Gray, *History of Agriculture in the Southern United States to 1860*, 2 vols. (Washington, D.C.: Carnegie Institution of Washington, 1933), vol. 1, 877.

20. Leonard W. Brinkmann, "The Historical Geography of Improved Cattle in the United States to 1870," Ph.D. dissertation, University of Wisconsin, 1964; Charles T. Leavitt, "Attempts to Improve Cattle Breeds in the United States, 1790–1860," *Agricultural History* 7 (1933), 51–61; Percy Wells Bidwell and John I. Falconer, *History of Agriculture in the Northern United States, 1620–1860* (Washington, D.C.: Carnegie Institution of Washington, 1925), 849; James W. Whitaker, *Feedlot Empire* (Ames: Iowa State University Press, 1975), 23; and Jonas Viles, "Sections and Sectionalism in a Border State," *Mississippi Valley Historical Review* 21 (1934–1935), 11.

21. Alexis de Tocqueville, *Democracy in America*, vol. 1 (New York: Vintage Books, 1990), 362. Charles T. Leavitt, "Transportation and the Livestock Industry of the Middle West to 1860," *Agricultural History* 8 (1934), 26, states, "from the viewpoint of maturity of agricultural development, the Middle West was divided into two districts, Ohio and Kentucky constituting one, and the region to the West and North the other."

22. Charles Wayland Towne and Edwin N. Wentworth, *Cattle and Men* (Norman: University of Oklahoma Press, 1955), 137–47; Gary S. Dunbar, "Colonial Carolina Cowpens," *Agricultural History* 35 (1961), 125–32; Frank Lawrence Owsley, *Plain Folk of the Old South* (Baton Rouge: Louisiana State University Press, 1949), 25–60; Terry G. Jordan, "Early Northeast Texas and the Evolution of Western Ranching," *Annals*, Association of American Geographers, 67 (1977), 66–87.

23. I. T. Frary, *Early Homes of Ohio* (Richmond, Va.: Garrett & Massie, 1936), 39; Mary Tolbert, Circleville, Ohio, personal communication.

24. Paul C. Henlein, *Cattle Kingdom in the Ohio Valley, 1783–1860* (Lexington: University Press of Kentucky, 1959), 10; also see Henlein, "Early Cattle Ranches of the Ohio Valley," *Agricultural History* 35 (1961), 150–54; and Robert Leslie Jones, "The Beef Cattle Industry in Ohio prior to the Civil War," *Ohio Historical Quarterly* 64 (1955), 168–94, 287–319.

25. Bidwell and Falconer, *Northern United States*, 388–89.

26. Renick, *Memoirs*, 11.

27. Richard Orr Curry, *A House Divided* (Pittsburgh: University of Pittsburgh Press, 1964); 53, 108–109. John J. Winberry, "Formation of the West Virginia-Virginia Boundary, *Southeastern Geographer* 17 (1977), 108–24, also suggests that the boundary between the two states was not a cultural divide.

28. Henlein, *Cattle Kingdom*, 2; E. I. Renick, "The Renick Family of Virginia," *Publications of the Southern History Association* 3 (1899), 221–27. For a geographical interpretation

see fig. 2, "Cultural Diffusion circa 1810," in Robert D. Mitchell, "The Formation of Early American Cultural Regions: An Interpretation," James R. Gibson, ed., *European Settlement and Development in North America* (Toronto: University of Toronto Press, 1976), 66–90; and fig. 4 in Richard Pillsbury, "The Pennsylvania Culture Area: A Reappraisal," *North American Culture* 3, no. 2 (1987), 37–54. A very different interpretation is given by David Hackett Fischer, *Albion's Seed: Four British Folkways in America* (New York: Oxford University Press, 1989), who regards "Backcountry" as a separate complex; Fischer's notions are examined critically in Wilbur Zelinsky's review of *Albion's Seed*; *Annals*, Association of American Geographers, 81 (1991), 526–31.

29. Henry Glassie, *Pattern in the Material Folk Culture of the Eastern United States* (Philadelphia: University of Pennsylvania Press, 1968); Fred B. Kniffen, "Folk Housing: Key to Diffusion," *Annals*, Association of American Geographers, 55 (1965), 549–77; Terry G. Jordan and Matti Kaups, *The American Backwoods Frontier* (Baltimore: Johns Hopkins University Press, 1989). Transformation of the Shenandoah Valley after immigration from the Tidewater is presented in Robert D. Mitchell, *Commercialism and Frontier* (Charlottesville: University of Virginia Press, 1979).

30. *People on the Farm: Corn and Hog Farming* (Washington, D.C.: U.S. Department of Agriculture, Office of Governmental and Public Affairs, 1977), 12; *The Mortgage Lifter, A Practical Hog Book* (Chicago: Hog Breeders, 1936).

6. Razorbacks and Poland-Chinas

1. Robert Leslie Jones, *History of Agriculture in Ohio to 1880* (Kent: Kent State University Press, 1983), chap. 1; and Jones, "Ohio Agriculture in History," *Ohio Historical Quarterly* 65 (1956), 229–58.

2. Martin R. Kaatz, "The Black Swamp: A Study in Historical Geography," *Annals*, Association of American Geographers, 45 (1955), 12–35.

3. Ohio State Board of Agriculture, *First Annual Report*, 1846; 50–51, 61–70.

4. S. Keener, "The Wheat Crop and Its Culture," Ohio Agriculture, *Third Annual Report*, 1848; 158–61.

5. See two articles in *Agricultural History* by Steven L. Stover: "Early Sheep Husbandry in Ohio," 36 (1962), 101–107; and "Ohio's Sheep Year: 1868," 38 (1964), 102–108. The worn-out lands could be restored, as Louis Bromfield showed (Charles E. Little, *Louis Bromfield at Malabar Farm* [Baltimore: Johns Hopkins University Press, 1988]).

6. Beverley W. Bond, Jr., *The Civilization of the Old Northwest: A Study of Political, Social, and Economic Development* (New York: Macmillan, 1934), 13.

7. Beverley W. Bond, Jr., *The Correspondence of John Cleves Symmes* (New York: Macmillan, 1926), 8; William Thomas Hutchinson, "Bounty Lands of the American Revolution in Ohio," Ph.D. dissertation, University of Chicago, 1927 (Arno Press Facsimile Edition, 1979), 91–92.

8. Lindsay Metcalfe Brien, *Miami Valley Records* [1935] (Miami Valley Genealogical Society, 1986); William E. Smith and Ophia D. Smith, *History of Southwestern Ohio, the Miami Valleys*, vol. 1 (New York: Lewis Historical, 1964), 158; Bernhard Knollenberg, *Pioneer Sketches of the Upper Whitewater Valley* (Indianapolis: Indiana Historical Society, 1945); James M. Berquist, "Tracing the Origins of Midwestern Culture," *Indiana Magazine of History* 77 (1981), 2–32.

9. John W. Reps, *Town Planning in Frontier America* (Columbia: University of Missouri Press, 1980), 201–203.

10. Bond, *Symmes*, 10–21.

11. Bond, *Old Northwest*, 362; Charles Cist, *Sketches and Statistics of Cincinnati in 1851* (Cincinnati: Wm. H. Moore, 1851), reports construction of 350-ton vessels continued at Marietta into the 1850s.

12. Pork receipts and shipments at New Orleans grew steadily after 1820 and reached a peak about 1855, when railroads began capturing the traffic in the north. See Sam B. Hilliard, "Antebellum Interregional Trade: The Mississippi River as an Example," in Ralph E. Ehrenberg, ed., *Pattern and Process: Research in Historical Geography* (Washington, D.C.: Howard University Press, 1975), 207.

13. Ohio Agriculture, *First Annual Report*, 1846, 48; Jones, "Ohio Agriculture," 236–37. The contributions of agriculture and industry to the urban system are studied in Edward K. Muller, "Selective Urban Growth in the Middle Ohio Valley, 1800–1860," *Geographical Review* 66 (1976), 178–99. On the early economic history of the Miami Valley see Wilfrid Gladstone Richards, "The Settlement of the Miami Valley of Southwestern Ohio," Ph.D. dissertation, University of Chicago, 1948; anecdotes of settlement farther north in the valley are found in Leonard U. Hill et al., *A History of Miami County, Ohio, 1807–1953* (Piqua, Ohio: County Sesquicentennial Organizing Committee, 1953).

14. Rudolph A. Clemen, *The American Livestock and Meat Industry* (New York: Ronald Press, 1923), 93; Charles T. Leavitt, "Transportation and the Livestock Industry of the Middle West to 1860," *Agricultural History* 8 (1934), 20; and Leavitt, "Some Aspects of the Western Meat Packing Industry, 1830–1860," *Journal of Business of the University of Chicago* 4 (1931), 68–90. Also see Richard Wade, *The Urban Frontier: The Rise of Western Cities, 1790–1830* (Cambridge: Harvard University Press, 1967), 322–26.

15. Cist, *Cincinnati in 1851*, 228, reports that a pack of 375,000 hogs per year was typical in the late 1840s. The Kentucky contribution is difficult to determine because "so much of that produced in Kentucky was sold from markets in Cincinnati"; Thomas D. Clark, "Livestock Trade between Kentucky and the South, 1840–1860," *Register of the Kentucky State Historical Society* 27 (1929), 576. Cincinnati's packing plants are considered a source of innovation; see Richard O. Arms, "From Disassembly to Assembly: Cincinnati, the Birthplace of Mass Production," *Business History Review* 45 (1971), 19–34. Meat packing was part of a system of production that has been seen as characterizing the whole region (Brian Page and Richard Walker, "From Settlement to Fordism: The Agro-Industrial Revolution in the American Midwest," *Economic Geography* 67 [1991], 281–315).

16. M. E. Ensminger, *Swine Science* (Danville, Ill.: Interstate, 1961), 1–2; J. L. Krider and W. E. Carroll, *Swine Production* (New York: McGraw-Hill, 1971), 921; and Merrill K. Bennett, "Aspects of the Pig," *Agricultural History* 44 (1970), 223–36.

17. Robert Beverley, *The History of Virginia* [1772] (Richmond: J. W. Randolph, 1855), 25. Carl O. Sauer, *Sixteenth Century North America: The Land and People as Seen by the Europeans* (Berkeley: University of California Press, 1971), 183–85, describes the introduction of hogs by de Soto.

18. Ensminger, *Swine Science*, 1–2; Sam B. Hilliard, *Hog Meat and Hoe Cake* (Carbondale: Southern Illinois University Press, 1972), 101.

19. Percy Wells Bidwell and John I. Falconer, *History of Agriculture in the Northern United States, 1620–1860* (Washington, D.C.: Carnegie Institution of Washington, 1925) 437–41. In the prairie-woodland setting of Illinois "hogs were taken in early fall to the heavy timber of the bottoms and then turned loose to fatten on the mast"; Hubert Schmidt, "Farming in Illinois a Century Ago as Illustrated in Bond County," *Journal of the Illinois State Historical Society* 31 (1938), 149. Frederick C. Minkler, *Pigs, Patriotism, and Profits* (New York: Advanced Agricultural, 1918), frontispiece, shows hogs in a woodland setting, with ear corn scattered on the ground, the "ideal conditions for pork production."

20. Earl B. Shaw, "Geography of Mast Feeding," *Economic Geography* 16 (1940), 233–49, shows that mast feeding was a specialty of the southern uplands in the United States; it was also a fixture in the cork-oak lands of Spain and Portugal. Soft, greasy pork and fluid lard result from a mast diet and are produced also when hogs are fed only peanuts; E. Z. Russell et al., "Hog Production and Marketing," *Yearbook of Agriculture, 1922* (Washington, D.C.: U.S. Department of Agriculture, 1922), 214–15.

21. Ensminger, *Swine Science*, 42–50; H. C. Dawson, *The Hog Book* (Chicago: Breeder's Gazette, 1913), 34–35; and Robert J. Evans, *History of the Duroc* (James J. Doty, 1918), 7.

22. Ensminger, *Swine Science*, 52; Dawson, *Hog Book*, 57; Joseph Ray Davis and Harvey S. Duncan, *History of the Poland-China Breed of Swine* (Maryville, Mo.: Poland China History Assoc., 1921), 1–5; Russell H. Anderson, "Agriculture among the Shakers, Chiefly at Mount Lebanon," *Agricultural History* 24 (1950), 116.

23. S. M. Shepard, *The Hog in America* (Indianapolis: Swine Breeders Journal, 1886), 10–18; Lewis Cecil Gray, *History of Agriculture in the Southern United States to 1860*, 2 vols. (Washington, D.C.: Carnegie Institution of Washington, 1933), vol. 1, 847–53.

24. Davis and Duncan, *Poland China Breed*, 11; Ensminger, *Swine Science*, 52–53.

25. "David Meeker Magie," *Memorial Record of Butler County, Ohio* (Chicago: Record, 1894), 395; George C. Crout, *Butler County: An Illustrated History* (Woodland Hills, Calif.: Windsor, 1984), 33–34. Dawson, *Hog Book*, 13, states: "Magie felt piqued by [the] selection of the name Poland-China instead of Magie hog. The result was ill-feeling for many years afterward." Both Magie and Poland China were well-known breeds and considered equivalent in Kansas and Nebraska at the time "Poland China" was made official; see F. Dwight Coburn, "Essay on Raising, Feeding, and Management of Hogs in Kansas," Transactions of the Kansas State Board of Agriculture, 1872, 276; and Irving L. Lyman, "Hogs in Nebraska," Nebraska State Board of Agriculture *Annual Report*, 1873, 275. The Illinois State Fair hog listings for 1869–1870 included "Poland, and China, and Magie breeds" (*Transactions of the Illinois State Agricultural Society* [Springfield, 1871]).

26. Davis and Duncan, *Poland China Breed*, 6–8.

27. Ohio Agriculture, *First Annual Report*, 1846, 56.

28. Thomas D. Clark, "Livestock Trade between Kentucky and the South, 1840–1860," *Register of the Kentucky State Historical Society* 27 (1929), 576; Gray, *Southern United States*, vol. 1, 841.

29. Clark, "Livestock Trade," 575; Gray, *Southern United States*, vol. 1, 854; Steven V. Ashby, *Middle Tennessee* (Baton Rouge: Louisiana State University Press, 1988), 19.

30. Eugene D. Genovese, "Livestock in the Slave Economy of the Old South—A Revised View," *Agricultural History* 36 (1962), 149.

31. Hilliard, *Hog Meat and Hoecake*, 95.

32. Genovese, "Livestock in the Slave Economy," 143–49.

33. Elizabeth L. Parr, "Kentucky's Overland Trade with the Antebellum South," *Filson Club History Quarterly* 2 (1928), 71–81; Charles T. Leavitt, "Transportation and the Livestock Industry of the Middle West to 1860," *Agricultural History* 8 (1934), 29; and Clark, "Livestock Trade," 569–71, give a detailed impression of droving in the South. Edmund Cody Burnett, "Hog Raising and Hog Driving in the Region of the French Broad River," *Agricultural History* 20 (1946), 86–103, describes the most important river followed. This water-level route through Knoxville and Asheville requires only a short, steep descent of the Blue Ridge beyond the Asheville Basin. Significantly, the Atlantic-Gulf drainage divide is projected farther east across the mountains here, in terms of physiographic provinces, than it is in the northern Appalachians (William D. Thornbury, *Regional Geomorphology of the United States* [New York: Wiley, 1965], 105).

34. Leavitt, "Livestock Industry," 29. Parr, "Kentucky's Overland Trade," 74, confirms large corn crops raised in the Appalachian valleys, which farmers "sold to Kentucky drovers at a sacrifice in the effort to get ready cash."

7. The First Corn Belt

1. Margaret Walsh, "Pork Packing as a Leading Edge of Midwestern Industry, 1835–1875," *Agricultural History* 51 (1977), 702–17; and Walsh, "The Spatial Evolution of the Midwestern Pork Industry, 1835–1875," *Journal of Historical Geography* 4 (1978), 1–22.

2. Quoted in John D. Barnhart and Dorothy L. Riker, *Indiana to 1816* (Indianapolis: Indiana Historical Bureau and Indiana Historical Society, 1971), 139. The early Wabash is described in Elbert J. Benton, *The Wabash Trade Route in the Development of the Old Northwest* (Baltimore: Johns Hopkins University Press, 1903).

3. On early Vincennes see R. Louis Gentilcore, "Vincennes and French Settlement in the Old Northwest," *Annals*, Association of American Geographers, 47 (1957), 285–97. Physical geography is described in Alden Cutshall, "Vincennes: Historic City on the Wabash," *Scientific Monthly* 57 (1943), 413. The first British expedition up the Mississippi observed some wheat or other small grains (their "corn") but usually found maize; see Captain Philip Pittman, *The Present State of the European Settlements on the Mississippi* [1770], Frank Heywood Hodder, ed. (Cleveland: Arthur H. Clark, 1906), 96, 102.

4. Barnhart and Riker, *Indiana*, 337–38.

5. Bil Gilbert, *God Gave Us This Country* (New York: Atheneum, 1989), 243; Barnhart and Riker, *Indiana*, 390n; Richard White, *The Middle Ground: Indians, Empires, and Republics in the Great Lakes Region, 1650–1815* (Cambridge: Cambridge University Press, 1991), 516.

6. Beverley W. Bond, Jr., *The Correspondence of John Cleves Symmes* (New York: Macmillan, 1926), 278–81.

7. John Reps, *Town Planning in Frontier America* (Columbia: University of Missouri Press, 1980), 279.

8. Morris Birkbeck, *Letters from Illinois* (Philadelphia: M. Carey and Son, 1818); George Flower, *History of the English Settlement in Edwards County, Illinois*, 2d ed., Chicago Historical Society Collections, vol. 1 (Chicago: Fergus, 1909).

9. Douglas R. McManis, *The Initial Evaluation and Utilization of the Illinois Prairies, 1815–1840* (Chicago: University of Chicago, Department of Geography, Research Paper 94, 1964), 72–76.

10. Leslie Hewes and Christian L. Jung, "Early Fencing on the Middle Western Prairie," *Annals*, Association of American Geographers, 71 (1981), 180.

11. McManis, *Illinois Prairies*, 75; Arthur C. Boggess, *The Settlement of Illinois, 1778–1803*, Chicago Historical Society Collections, vol. 5 (Chicago, 1908), 154.

12. B. P. Birch, "The Environment and Settlement of the Prairie-Woodland Transition Belt—A Case Study of Edwards County, Illinois," *Southampton Research Series in Geography*, no. 6 (1971), 3–30. Also see Wayne E. Kiefer, *Rush County, Indiana: A Study in Rural Settlement Geography*, Monograph no. 2 (Bloomington: Indiana University Department of Geography, 1969).

13. John Woods, *Two Years' Residence in the Settlement on the English Prairie, in the Illinois Country, United States* (London: Longman, Hurst, Rees, Orme, and Brown, 1822), 213.

14. John M. Peck, *A Gazetteer of Illinois* (Jacksonville, Ill.: R. Goudy, 1834), summarized in McManis, *Illinois Prairies*, 83–85.

15. Edgar N. Transeau, "The Prairie Peninsula," *Ecology* 16 (1935), 423–37, remains the standard "either-or" map of the middle-western prairie. Transeau included a few bottom-lands, such as the Pickaway Plains, in the prairie category.

16. McManis, *Illinois Prairies*, 25–27.

17. C. Vann Woodward, *Origins of the New South* (Baton Rouge: Louisiana State University Press, 1951), 109–10.

18. John Carroll Power, *History of the Early Settlers of Sangamon County, Illinois* (Springfield: Edwin A. Wilson, 1876).

19. Donald W. Meinig, *The Shaping of America*, vol. 1 of *Atlantic America, 1492–1800* (New Haven: Yale University Press, 1986), figs. 29, 31.; Robert D. Mitchell, "The Presbyterian Church as an Indicator of Westward Expansion in Eighteenth Century America," *Professional Geographer* 18 (1966), 293–99.

20. John Mack Faragher, *Sugar Creek* (New Haven: Yale University Press, 1986), 123, states, "nine out of ten Sugar Creek [rural Sangamon County] settlers before 1840 were from the South."

21. Roger D. Mason, *Euro-American Settlement Systems in the Central Salt River Valley of Northeast Missouri*, vol. 1, Cannon Reservoir Human Ecology Project (Columbia: University of Missouri, Department of Anthropology, 1984).

22. Birkbeck, *Letters from Illinois*, 68. Jeremy Atack and Fred Bateman, "Yankee Farming and Settlement in the Old Northwest: A Comparative Analysis," in David C. Klingaman and Richard K. Vedder, eds., *Essays on the Economy of the Old Northwest* (Athens: Ohio University Press, 1987), 85. Atack and Bateman were countering Lois Kimball Mathews, *The Expansion of New England* (New York: Russell and Russell, 1909), 208.

23. Mary Vose Harris, "The Autobiography of Benjamin Franklin Harris," *Transactions of the Illinois Historical Society* (Springfield: Illinois State Historical Society, 1923), 72–101.

24. Richard Bardolph, "Illinois Agriculture in Transition, 1820–1870," *Journal of the Illinois State Historical Society* 41 (1948), 250, 417.

25. Joseph G. McCoy, *Historic Sketches of the Cattle Trade of the West and Southwest* [1874] (Ann Arbor: University Microfilms, 1966), 169–78; cf. William Renick, "On the Cattle Trade of the Ohio Valley," Ohio State Board of Agriculture, *Third Annual Report*, 1848 (Columbus, 1849), 162–64.

26. Clarence P. McClelland, "Jacob Strawn and John T. Alexander, Central Illinois Stockmen," *Illinois State Historical Society Journal* 34 (1941), 177–208; and Charles M. Eames, *Historic Morgan and Classic Jacksonville* (Jacksonville, Ill.: Daily Steam Job Printing Office, 1885). At least 23,109 cattle, 50,720 hogs, and 109,330 bushels of wheat were reported shipped from Morgan County by rail in 1861; *Transactions of the Illinois State Agricultural Society* 5 (Springfield, 1865).

27. Helen M. Cavanagh, *Funk of Funk's Grove: Farmer, Legislator, and Cattle King of the Old Northwest, 1797–1865* (Bloomington, Ill.: Pantagraph, 1952), chap. 1.

28. Cavanagh, *Funk's Grove*, 46–67.

29. Margaret Beattie Bogue, *Patterns from the Sod* (Springfield: Illinois State Historical Library, 1959), 48.

30. Bogue, *Patterns from the Sod*, 76.

31. Leslie Hewes, "The Northern Wet Prairie of the United States: Nature, Sources of Information, and Extent," *Annals*, Association of American Geographers, 41 (1951), 307–23; and Margaret Beattie Bogue, "The Swamp Land Act and Wet Land Utilization in Illinois, 1850–1890," *Agricultural History* 25 (1951), 169–80. Roger A. Winsor, "Environmental Imagery of the Wet Prairie of East Central Illinois, 1820–1870," *Journal of Historical Geography* 13 (1987), 375–97, claims the prairie was not as wet as portrayed, but rather (394) "suffered from adverse spatial stereotyping."

32. Sherman N. Geary, "The Cattle Industry of Benton County," *Indiana Magazine of History* 21 (1925), 27–32; Paul W. Gates, "Hoosier Cattle Kings," *Indiana Magazine of History* 44 (1948), 1–24.

33. The sequence on some of the wetter lands is portrayed by Alfred H. Meyer, "The Kankakee 'Marsh' of Northern Indiana and Illinois," in Robert S. Platt, ed., *Field Study in American Geography*, Research Paper no. 61 (Chicago: University of Chicago, Department of Geography, 1959), 200–16.

34. Imports between 1848 and 1886 are listed in Donald R. Ornduff, *The Hereford in America* (Kansas City: by author, 1957), 64–66; Geary, "Benton County," 31.

35. Gates, "Cattle Kings," 2. This contention is part of Gates's larger argument about the significance of frontier landlords. See Gates, "Frontier Estate Builders and Farm Laborers," in Walker D. Wyman and Clifton B. Kroeber, eds., *The Frontier in Perspective* (Madison: University of Wisconsin Press, 1957), 144–63.

36. *Counties of Warren, Jasper, Benton, and Newton, Indiana; Historical and Biographical* (Chicago: F. A. Battey, 1883).

37. For example, see *Portrait and Biographical Record of Iroquois County, Illinois* (Chicago: Lake City, 1893), and *The History of McLean County, Illinois* (Chicago: Wm. Le Baron Jr., 1879).

38. Mildred Throne, "Southern Iowa Agriculture, 1833–1890," *Agricultural History* 23 (1949), 124–30.

39. *Report and Proceedings of the Iowa State Agricultural Society*, 1855 (Fairfield, Ia., 1856), 241.

40. Theodore L. Carlson, *The Illinois Military Tract* (Urbana: University of Illinois Press, 1951), 26.

41. Bardolph, "Illinois Agriculture," 430.

42. On the association of barn types with settlement groups, see Allen G. Noble, *Wood, Brick, and Stone: The North American Settlement Landscape*, 2 vols. (Amherst: University of Massachusetts Press, 1984). Pennsylvania's contributions are described in Robert F. Ensminger, *The Pennsylvania Barn: Its Origin, Evolution, and Distribution* (Baltimore: Johns Hopkins University Press, 1992); and Alfred L. Shoemaker, ed., *The Pennsylvania Barn* (Kutztown: Pennsylvania Folklife Society, 1959).

43. Divisions in terms of migrant origins on the map are generalized from maps based on birthplaces recorded in county histories; see John C. Hudson, "North American Origins of Middlewestern Frontier Populations," *Annals*, Association of American Geographers, 78 (1988), 395–413.

44. Randolph Chandler Downes, *Frontier Ohio, 1788–1803* (Columbus: Ohio State Archaeological and Historical Society, 1935), 113–26.

45. Margaret Walsh, *The Rise of the Midwestern Meat Packing Industry* (Lexington: University Press of Kentucky, 1982), 36–37.

46. Meat-packing data taken from *Report of the U.S. Commissioner of Patents for the Year 1849*, part 2, Agriculture (Washington, D.C., 1850), and Thomas D. Clark, "Livestock Trade between Kentucky and the South, 1840–1860," *Register of the Kentucky State Historical Society* 27 (1929), 569–81. See also the series of maps in Walsh, *Midwestern Meat Packing*.

47. Charles T. Leavitt, "Transportation and the Livestock Industry of the Middle West to 1860," *Agricultural History* 8 (1934), 22.

48. The rapid transition from steamboats to railroads is described in Timothy F. Mahoney, *River Towns in the Great West: The Structure of Provincial Urbanization in the American Midwest, 1820–1870* (Cambridge: Cambridge University Press, 1990), 2–9, 248. Also see Wyatt B. Belcher, *The Economic Rivalry between Chicago and St. Louis* (New York: Columbia University Press, 1947).

8. Corn Belt Sectionalism

1. Malcolm Rohrbaugh, "Diversity and Unity in the Old Northwest, 1790–1850: Several Peoples Fashion a Single Region," *Pathways to the Old Northwest* (Indianapolis: Indiana Historical Society, 1988), 76.

2. Richard Lyle Power, *Planting Corn Belt Culture: The Impress of the Upland Southerner and Yankee in the Old Northwest*, Indiana Historical Society Publications, no. 25 (Indianapolis: Indiana Historical Society, 1953), 172.

3. William Vipond Pooley, *The Settlement of Illinois from 1830 to 1850*, Bulletin of the University of Wisconsin, no. 1 (Madison: University of Wisconsin Press, 1908), 287–595.

4. Henry C. Hubbart, " 'Pro-Southern' Influences in the Free West, 1840–1865," *Mississippi Valley Historical Review* 20 (1933), 47.

5. William Carter, *Middle West Country* (Boston: Houghton-Mifflin, 1975), 38.

6. Lewis Cecil Gray, *History of Agriculture in the Southern United States to 1860*, 2 vols. (Washington, D.C.: Carnegie Institution of Washington, 1933).

7. Robert E. Chaddock, *Ohio before 1850: A Study of the Early Influence of Pennsylvania and Southern Populations in Ohio*, Columbia University Studies in History, Economics and Public Law, vol. 21, no. 2 (New York: Columbia University Press, 1908), chap. 4.

8. John D. Barnhart and Dorothy L. Riker, *Indiana to 1816: The Colonial Period* (Indianapolis: Indiana Historical Bureau and Indiana Historical Society, 1971), 315–47. Petitions favoring slavery are included in Jacob Piatt Dunn, Jr., "Slavery Petitions and Papers," *Indiana State Historical Society Publications*, no. 2 (Indianapolis: Indiana Historical Society, 1894), 443–529. On early migration to Indiana see John D. Barnhart, "The Southern Influence in the Formation of Indiana," *Indiana Magazine of History* 33 (1937), 261–76; and Barnhart, "The Migration of Kentuckians across the Ohio River," *Filson Club History Quarterly* 25 (1951), 24–32. Also see Gregory S. Rose, "Major Sources of Indiana's Settlers in 1850," *Pioneer America Society Transactions* 6 (1983), 67–76; Rose, "Hoosier Origins: The Nativity of Indiana's United States-Born Population in 1850," *Indiana Magazine of History* 81 (1985), 201–32; and Rose, "Upland Southerners: The County Origins of Southern Migrants to Indiana by 1850," *Indiana Magazine of History* 82 (1986), 242–63.

9. Arthur C. Boggess, *The Settlement of Illinois, 1778–1803*, Collections, vol. 5 (Chicago: Chicago Historical Society, 1908), chap. 6. The sources of migration to Illinois are discussed in John D. Barnhart, "The Southern Influence in the Formation of Illinois," *Illinois State Historical Society Journal* 32 (1939), 358–79; and Douglas K. Meyer, "Southern Illinois Migration Fields: The Shawnee Hills in 1850," *Professional Geographer* 28 (1976), 151–60.

10. For a different view of the threat of slavery to the Corn Belt, see Carville V. Earle, "A Staple Interpretation of Slavery and Free Labor," *Geographical Review* 68 (1978), 51–65; and Earle, "Regional Economic Development West of the Appalachians, 1815–1860," in Robert D. Mitchell and Paul A. Groves, eds., *North America* (Totowa, N.J.: Rowman & Littlefield, 1987), 172–97. Earle's 1978 paper has been republished in Earle, *Geographical Inquiry and American Historical Problems* (Stanford: Stanford University Press, 1992), 226–57.

11. Boggess, *Settlement of Illinois*, 179; Edgar Allan Holt, *Party Politics in Ohio, 1840–1850*, Ohio Historical Collections, vol. 1 (Columbus: Ohio State Archaeological and Historical Society, 1931), 244; Chaddock, *Ohio before 1850*, 83.

12. Leonard Erickson, "Politics and Repeal of Ohio's Black Laws, 1837–1849," *Ohio History* 82 (1983), 154–75; and Elmer Gertz, "The Black Laws of Illinois," *Journal of the Illinois State Historical Society* 56 (1963), 454–73.

13. Boggess, *Settlement of Illinois*, 180–81.

14. Earle, "Slavery and Free Labor," 61–62; Earle, "Regional Economic Development," 190–91.

15. *Journals of the Continental Congress* 25 (1783), 560–61.

16. Harrison Anthony Trexler, *Slavery in Missouri, 1804–65* (Baltimore: Johns Hopkins University Press, 1914), 58–59, 103; and Perry McCandless, *A History of Missouri*, vol. 2, 1820 to 1860 (Columbia: University of Missouri Press, 1972), 57.

17. David Brion Davis, "The Significance of Excluding Slavery from the Old Northwest in 1787," *Indiana Magazine of History* 84 (1988), 75–89. The full context of the Ordinances is presented in Peter S. Onuf, *Statehood and Union: A History of the Northwest Ordinance* (Bloomington: Indiana University Press, 1987).

18. Carl O. Sauer, *The Geography of the Ozark Highland of Missouri* [1920] (New York: AMS Press, 1971), 109; and Walter A. Schroeder, "Spread of Settlement in Howard County, Missouri, 1810–1859," *Missouri Historical Review* 62 (1968), 1–37. Schroeder states (2–3) that "the Boonslick was forested" and that this "familiar forest environment," plus deep, loessial soils made the Boonslick the first choice of settlers. For a detailed analysis of the early Boonslick settlement, see Michael J. O'Brien, "The Roots of Frontier Expansion," in O'Brien et al., *Grassland, Forest, and Historical Settlement* (Lincoln: University of Nebraska Press, 1984), 74–94. The geographical context of early Missouri settlement is provided in James R. Shortridge, "The Expansion of the Settlement Frontier in Missouri," *Missouri Historical Review* 75 (1980), 64–90.

19. On environmental conditions and Missouri slavery, see James Fernando Ellis, *The Influence of Environment on the Settlement of Missouri* (St. Louis: Webster, 1929), 110; and William Wilson Elwang, *The Negroes of Columbia, Missouri* (Columbia: University of Missouri, Department of Sociology, 1904). Slavery and hemp in Kentucky and Missouri are discussed in James F. Hopkins, *The Hemp Industry in Kentucky* (Lexington: University Press of Kentucky, 1951), 135–37.

20. Jonas Viles, "Sections and Sectionalism in a Border State," *Mississippi Valley Historical Review* 21 (1934–1935), 3–22.

21. McCandless, *History of Missouri*, vol. 2, 60–62. The flurry of abolitionist sentiment in the 1830s soon died out. For an opposing view on the status of Missouri slavery in the late antebellum period, see Walter Warrington Ryle, *Missouri: Union or Secession* (Nashville: George Peabody College for Teachers, 1931).

22. Trexler, *Slavery in Missouri*, 22.

23. Ralph V. Anderson and Robert E. Gallman, "Slaves as Fixed Capital: Slave Labor and Southern Economic Development," *Journal of American History* 64 (1977), 38.

24. Elizabeth L. Parr, "Kentucky's Overland Trade with the Antebellum South," *Filson Club History Quarterly* 2 (1928), 73, states that many Kentuckians were employed as drovers. While this was undoubtedly true, in the census of 1850 only 29 Kentuckians listed "drover" as their occupation; among the Corn Belt states droving could have been considered as a primary occupation only in Ohio, which reported 340 drovers. Most drovers evidently did other work and probably many were farmers or seasonal farm workers. See David E. Schob, *Hired Hands and Plowboys: Farm Labor in the Midwest, 1815–1860* (Urbana: University of Illinois Press, 1975); Earle, "Regional Economic Development."

25. Eugene Berwanger, *The Frontier against Slavery* (Urbana: University of Illinois Press, 1967), 140, 111–120; Jacque Voegeli, "The Northwest and the Race Issue, 1861–1862," *Mississippi Valley Historical Review* 40 (1963), 235–36, 244–47. Also see Leon Litwack, *North of Slavery* (Chicago: University of Chicago Press, 1961).

26. Earle, "Slavery and Free Labor," 52.

27. Berwanger, *Frontier against Slavery*, 35.

28. Theodore Clark Smith, *Parties and Slavery, 1850–1859* (New York: Harper, 1907), 121.

29. Eric Foner, *Free Soil, Free Labor, Free Men: The Ideology of the Republican Party before the Civil War* (New York: Oxford University Press, 1970), 155–62. Chapter 5 in Andrew R. L. Cayton and Peter S. Onuf's *The Midwest and the Nation* (Bloomington: Indiana University Press, 1990), is an excellent introduction to the enormous literature on this subject.

30. Data for the voting maps were taken from Walter Dean Burnham, *Presidential Ballots* (Baltimore: Johns Hopkins University Press, 1955). The earliest analysis of maps like these was Frederick Jackson Turner, "Is Sectionalism in America Dying Away?" *American Journal of Sociology* 13 (1908), 661–75. A contemporary political-geographic viewpoint is provided in Fred M. Shelley and J. Clark Archer, "Sectionalism and Presidential Politics: Voting Patterns in Illinois, Indiana, and Ohio," *Journal of Interdisciplinary History* 20 (1989), 227–55.

31. See, for example, Stephen E. Maizlish, *The Triumph of Sectionalism* (Kent: Kent State University Press, 1983).

32. Charles A. Church, *History of the Republican Party in Illinois, 1854–1912* (Rockford, Ill.: Wilson, 1912), 20–32; and *Transactions of the McLean County Historical Society*, vol. 3 (Bloomington: Pantagraph, 1900), 45.

33. Geographical variation in German support for political parties at this time was complex; see the papers in Frederick C. Luebke, ed., *Ethnic Voters and the Election of Lincoln* (Lincoln: University of Nebraska Press, 1971). Locations of several dozen antiwar protests and Copperhead strongholds are mentioned in Frank L. Klement, *The Copperheads in the Middle West* (Chicago: University of Chicago Press, 1960); while the list presumably is not exhaustive, outside of immigrant concentrations such as eastern Wisconsin, most of the incidents identified occurred in Corn Belt counties.

34. Maps of the presidential vote, 1932–1968, may be found in John Fraser Hart, "The Middle West," *Annals*, Association of American Geographers, 62 (1972), 279. See also John H. Fenton, *Midwest Politics* (New York: Holt, Rinehart, and Winston, 1966); and J. Clark Archer et al., "The Geography of U.S. Presidential Elections," *Scientific American* 259 (1988), 18–25.

35. John C. Hudson, "North American Origins of Middlewestern Frontier Populations," *Annals*, Association of American Geographers, 78 (1988), fig. 1; Hudson, "Who Was Forest Man? Sources of Migration to the Plains," *Great Plains Quarterly* 6 (1986), 69–83.

9. Specialization and Westward Expansion

1. Two excellent histories of Chicago, taking different viewpoints, are Bessie Louise Pierce, *A History of Chicago*, 3 vols. (New York: Knopf, 1937–1957); and William Cronon, *Nature's Metropolis: Chicago and the Great West* (New York: Norton, 1991).

2. Michael P. Conzen and Melissa J. Morales, eds., *Settling the Upper Illinois Valley: Patterns of Change in the I&M Canal Corridor, 1830–1900*. Studies on the Illinois & Michigan Canal Corridor, no. 3 (Chicago: University of Chicago, Committee on Geographical Studies: 1989).

3. William T. Hutchinson, *Cyrus Hall McCormick, Seed Time, 1809–1856* (New York: Century, 1930), 252–53.

4. Rudolph A. Clemen, *The American Livestock and Meat Industry* (New York: Ron-

ald Press, 1923), 83–109; Louise Carroll Wade, *Chicago's Pride: The Stockyards, Packing-town and Environs in the Nineteenth Century* (Urbana: University of Illinois Press, 1987), chap. 3.

5. *Drovers Journal Yearbook of Figures*, no. 54 (Chicago: Corn Belt, 1954); *Transactions of the Illinois State Agricultural Society* 6 (Springfield, 1868), 314–31; origins are inferred from the tabulations by railroad companies and the location of lines of those companies.

6. *Report of Manufacturing Industries in the United States at the Eleventh Census*, 1890, part 2 (statistics of wholesale slaughtering and meat packing in 165 cities). The value of products minus the cost of materials shows Chicago's share to be $29.4 million of the total $59.9 million among all cities.

7. Clemen, *Livestock and Meat Industry*, 109; Illinois Transactions 6 (Springfield, 1868), 314–31.

8. Mary Yeager, *Competition and Regulation: The Development of Oligopoly in the Meat Packing Industry* (Greenwich, Conn.: JAI Press, 1981), 77.

9. Clemen, *Livestock and Meat Industry*, 150–54; Wade, *Chicago's Pride*, 64; Louis B. Swift and Arthur Van Vlisingen, *The Yankee of the Yards: The Biography of Gustavus Franklin Swift* (New York: A. W. Shaw, 1927); and Harper Leech and John C. Carroll, *Armour and His Times* (New York: Appleton-Century, 1938). Morrell was established in the western Corn Belt; see Lawrence Oakley Cheever, *The House of Morrell* (Cedar Rapids, Ia.: Torch Press, 1948).

10. Wade, *Chicago's Pride*, 202; Clemen, *Livestock and Meat Industry*, 159–62.

11. Mary Yeager Kujovich, "The Refrigerator Car and the Growth of the American Dressed Beef Industry," *Business History Review* 44 (1970), 460–82; Clemen, *Livestock and Meat Industry*, chap. 11; Yeager, *Competition and Regulation*, 49–67; Oscar E. Anderson, *Refrigeration in America* (Princeton: Princeton University Press, 1953).

12. Robert M. Aduddell and Louis P. Cain, "Location and Collusion in the Meat Packing Industry," Cain and Paul J. Uselding, eds., *Business Enterprise and Economic Change* (Kent, Ohio: Kent State University Press, 1973), 105–106. The refrigerator cars delivered meat to company-owned warehouses in major cities, and thus packers were largely able to eliminate wholesalers from the sale of processed goods (Glenn Porter and Harold C. Livesay, *Merchants and Manufacturers: Studies in the Changing Structure of Nineteenth Century Marketing* [Baltimore: Johns Hopkins University Press, 1971], 168–73).

13. Cronon, *Nature's Metropolis*, 114; and Jonathan Lurie, *The Chicago Board of Trade, 1859–1905: The Dynamics of Self-Regulation* (Urbana: University of Illinois Press, 1979).

14. Illinois *Transactions* 3 (1857–1858), 305–309, 518–26; 5 (1861–1864), 361, 514–19.

15. Shigehiro Yuasa, "The Commercial Pattern of the Illinois & Michigan Canal, 1848–1860," Conzen and Morales, eds., *Upper Illinois Valley*, 9–21, fig. 2; Illinois *Transactions* 6, 412–15; John G. Clark, *The Grain Trade in the Old Northwest* (Urbana: University of Illinois Press, 1966), 89–93.

16. The role of water transportation in stimulating Chicago's cash grain business does not suggest, however, that more canals might have substituted for railroads, as was argued in Robert W. Fogel, *Railroads and American Economic Growth: Essays in Econometric History* (Baltimore: Johns Hopkins University Press, 1964), 219. Fogel's hypothetical new canals (his fig. A-1) assumed the existing drainage of the land, and that would have made the Grand Prairie tributary to New Orleans!

17. *Transactions of the Department of Agriculture of the State of Illinois*, n.s., no. 1 (Springfield, 1872), 184, 321.

18. Transactions, *Agriculture of Illinois* 5 (1876), 177, 184, 232.

19. Quoted in Helen M. Cavanagh, *Seed, Soil and Science: The Story of Eugene D. Funk* (Chicago: Lakeside Press, 1959), 126.

20. Margaret Beattie Bogue, *Patterns from the Sod: Land Use and Land Tenure in the Grand Prairie, 1850–1900* (Springfield: Illinois State Historical Library, 1959), 146; Transactions, *Agriculture of Illinois* 12 (1882), 483–89.

21. That tenancy retarded growth in infrastructure was understood in Madison County, Ohio (Ohio State Board of Agriculture, *Third Annual Report*, 1848 [Columbus, 1849], 92). A decline in absentee ownership, and hence a decline in tenancy, in Ford County, Illinois, was noted favorably in 1876; *Transactions*, Agriculture of Illinois 5 (1876), 184.

22. Homer Socolofsky, *Landlord William Scully* (Lawrence: Regents Press of Kansas, 1979), 83–85. Although he was familiar with drainage in Ireland, Scully was not a drainage innovator in Illinois. Scully also followed the Corn Belt west, purchasing land in southeastern Nebraska and in Kansas; in 1888 he owned 182,894 acres in the United States (77).

23. John W. Alexander, "Freight Rates as a Geographic Factor in Illinois," *Economic Geography* 20 (1944), 25–30; John J. Hidore, "The Relationship between Cash-Grain Farming and Landforms," *Economic Geography* 39 (1963), 84–89; and Leverett P. Hoag, "Location Determinants for Cash-Grain Farming in the Corn Belt," *Professional Geographer* 14 (1962), 1–7. Hoag reviewed 14 reasons why Corn Belt farmers might market grain rather than feed it, and he concluded that cash-grain farming occurred on flat land because it was less subject to soil erosion when continuously planted to row crops. Also see Arlin D. Fentem, "Cash Feed Grain in the Corn Belt," Ph.D. dissertation, University of Wisconsin-Madison, 1974.

24. *Report and Proceedings of the Iowa State Agricultural Society*, 1855 (Fairfield, Ia., 1856), 214.

25. *Fifth Annual Report of the Iowa State Agricultural Society*, 1858; 439.

26. *History of Greene County, Missouri* (St. Louis: Western Historical, 1883); *The History of Jasper County, Missouri* (Des Moines: Mills, 1883); and *The History of Cass and Bates Counties, Missouri* (St. Joseph: National Historical, 1883).

27. Joseph G. McCoy, *Cattle Trade of the West and Southwest* [1874] (Ann Arbor: University Microfilms, 1966), 20–22.

28. Shorthorns had been brought to Kansas from Ohio in the 1860s; Charles Wood, "Upbreeding Western Range Cattle: Notes on Kansas, 1880–1920," *Journal of the West* 16 (1977), 17–18; see also Wood, *The Kansas Beef Industry* (Lawrence: Regents Press of Kansas, 1980).

29. Walter Prescott Webb, *The Great Plains* (Boston: Ginn, 1931), 212.

30. John E. Rouse, *The Criollo: Spanish Cattle in the Americas* (Norman: University of Oklahoma Press, 1977), 192; Marlene Felius, *Genus Bos* (Rahway, N.J.: Merck, 1985), 171.

31. Robert Dykstra, *The Cattle Towns* (New York: Knopf, 1968), 16; and W. Theodore Mealor, Jr., and Merle C. Prunty, "Open-Range Ranching in Southern Florida," *Annals*, Association of American Geographers, 66 (1976), 360–76.

32. McCoy, *Cattle Trade of the West*, 20, 169.

33. Bogue, *Patterns from the Sod*, 58–59.

34. John Carroll Power, *History of the Early Settlers of Sangamon County, Illinois* (Springfield: Edwin A. Wilson, 1876), 487–88.

35. Edwin H. Van Patten, "A Brief History of David McCoy and Family," *Journal of the Illinois State Historical Society* 14 (1921), 122–27.

36. McCoy, *Cattle Trade of the West*, 176. Joe McCoy was so feedlot oriented that he chided a Great Plains cattleman for not running hogs behind his cattle (discussed in Fred A. Shannon, *The Farmer's Last Frontier: Agriculture, 1860–1897* [New York: Harper Torchbook, 1968], 165).

37. McCoy, *Cattle Trade of the West*, 252–53.

38. Texas cattle reached as far north as southwestern North Dakota but were less important in terms of numbers than English cattle, which grazed in northern North Dakota (H. A. Pulling, "History of the Range Cattle Industry of Dakota," *South Dakota Historical Collections* 20 [1940], 47; and Ernest S. Osgood, *The Day of the Cattleman* [Minneapolis: University of Minnesota Press, 1929], 211). For northern North Dakota see Larry A. McFarlane, "British Remittance Men as Ranchers: The Case of Coutts Marjoribanks and Edmund Thursby, 1884–95," *Great Plains Quarterly* 11 (1991), 53–69.

39. McCoy, *Cattle Trade of the West*, 169, 237; Wood, "Upbreeding Western Cattle," 17.

40. Alvin Howard Sanders, *A History of Aberdeen Angus Cattle* (Chicago: Lakeside Press, 1928), chaps. 1–3; Donald R. Ornduff, *The Hereford in America* (Kansas City: by author, 1957), 64–66, chap. 11; Rouse, *The Criollo*, 193; and E. Heath-Agnew, *A History of Hereford Cattle and Their Breeders* (London: Gerald Duckworth, 1983).

41. Walter M. Kollmorgen and David S. Simonett, "Grazing Operations in the Flint Hills-Bluestem Pastures of Chase County, Kansas," *Annals*, Association of American Geographers, 55 (1965), 260–90, show that not all the acreage kept in pasture has uncultivable soils. Corn Belt livestock feeders came to the Flint Hills at the same time that English and Scottish interests were buying large tracts of pasture (Kollmorgen and Simonett, 290). Also see *Chase County Historical Sketches*, 2 vols. (Cottonwood Falls, Kans.: Chase County Historical Society, 1940, 1949).

42. James C. Malin, "An Introduction to the History of the Bluestem-Pasture Region of Kansas," *Kansas Historical Quarterly* 11 (1942), 6; also see Tom Isern, "Farmers, Ranchers, and Stockmen of the Flint Hills," *Western Historical Quarterly* 16 (1985), 253–64. The role of fire in maintaining the Flint Hills bluestem pastures was considered to be a deleterious one at the time Kollmorgen and Simonett wrote (287–88), but is now recognized as beneficial, even necessary; James Hoy, "Controlled Pasture Burning in the Folklife of the Kansas Flint Hills," *Great Plains Quarterly* 9 (1989), 231–38.

43. Atchison, Topeka and Santa Fe Railway, *Land Department Records*, Ledger Books, 1871–1874, Kansas State Historical Society, Center for Historical Research, Topeka. A sample of 300 persons who entered into land contracts with the railroad's land department, 1871–1874, showed that approximately half of them were living in the Grand Prairie of Illinois at the time; others lived primarily in the Corn Belt counties of Ohio, Indiana, or the remainder of Illinois.

44. *Transactions of the Kansas State Board of Agriculture*, Second Annual Report, 1873 (Topeka, 1874), 132, 169.

45. *Transactions*, Kansas Board of Agriculture, 1874 (Topeka, 1875), 82.

46. Carl B. Schmidt, "Reminiscences of Foreign Immigration Work for Kansas," *Transactions*, Kansas State Historical Society, 9 (1906), 485–97; and H. Craig Miner, *West of Wichita: Settling the High Plains of Kansas, 1865–1890* (Lawrence: University Press of Kansas, 1986), 80–92.

47. James C. Malin, *Winter Wheat in the Golden Belt of Kansas: A Study in Adaptation to Subhumid Geographical Environment* (Lawrence: University of Kansas Press, 1944), 250, wrote: "So far as the Mennonites were concerned, their contribution falls largely into the category of the accidents of history and there is no evidence yet available to demonstrate that they understood even remotely at the time the significance of what they were doing." Miner, *West of Wichita*, 82, doubts the Mennonite influence in wheat, stating "the trend toward winter wheat was already clear."

48. Annual statistics on crop production in the state are found in *The Annals of Kansas*, 1886–1925 (Topeka: Kansas State Historical Society, n.d.); Miner, *West of Wichita*, 120–21.

49. *Fifth Annual Report of the State Board of Agriculture* [Kansas] (Topeka, 1876), 490. The benefits and drawbacks of wheat monoculture versus diversified farming are illustrated in Aidan McQuillan, *Prevailing over Time: Ethnic Adjustment on the Kansas Prairies, 1875–1925* (Lincoln: University of Nebraska Press, 1990), see especially appendix D.

50. Second and third *Annual Report*, State Board of Agriculture [Nebraska] (Lincoln, 1869, 1871).

10. New Crops and Northward Expansion

1. Charles Foster, "The Hog," *Fifth Annual Report of the Iowa State Agricultural Society*, 1858; 147.

2. *Fifth Annual Report*, Iowa; 306.

3. *Fifth Annual Report*, Iowa; 221.

4. George F. Will, *Corn for the Northwest* (St. Paul: Webb, 1930), 27. Also see George F. Will and George E. Hyde, *Corn among the Indians of the Upper Missouri* (Lincoln: University of Nebraska Press, 1964).

5. *U.S. Census*, 1900, Agriculture, plate 7; the category "40 bushels per acre and over" includes counties in New England and northern Wisconsin where acreages were small but yields apparently were large.

6. George I. Quimby, *Indian Life in the Upper Great Lakes, 11,000 B.C. to A.D. 1800* (Chicago: University of Chicago Press, 1960), 73, wrote: "It seems probable that in Hopewell times [100 B.C.] the tropical flint corn had not yet been adapted to growth in cooler regions. Yet by A.D. 1700 a hardier Indian corn was being raised on the south side of Lake Superior, well north of the zone of Hopewell occupancy." D. W. Moodie and Barry Kaye, in "The Northern Limit of Indian Agriculture in North America," *Geographical Review* 59 (1969), 513–29, argue that the maximum northward penetration of Indian corn west of Lake Superior took place in the early decades of the nineteenth century.

7. Donald F. Jones, "Changes in Hybrid Seed Corn Production in the Future," *Progress in Corn Production* (American Seed Trade Association, 1949), 11. Corn grown under shade-tobacco cloth in Connecticut was taller.

8. "Experiments in the Corn Field," *First Annual Report of the Kansas Experiment Station* (Manhattan, 1888), 24.

9. William D. Emerson, *History and Incidents of Indian Corn and Its Culture* (Cincinnati: Wrightson, 1878), 165. Emerson relied on the U.S. Commissioner of Patents data published in the late 1840s; cf. Annual Reports, *Commissioner of Patents*, 1844–1850.

10. B. F. Johnson, "Indian Corn, Its Varieties, Culture, and Most Profitable Uses," *Transactions of the Department of Agriculture of the State of Illinois* 5, 1876, 121. In fact, daylength increases at an increasing rate with movement north in latitude in summer. On June 21, when the sun is overhead at noon at the Tropic of Cancer and every point north of 66.5° has 24 hours of daylight, the daylength, D, in hours at latitude, L, $0 < L < 66.5°$N, can be shown to be $D = 31.958 – 19.958 \cos L$. Daylength increases approximately two hours moving from latitude 30° to 40°, but by 2.45 hours moving from latitude 40° to 50°. By a similar argument, intensity of sunlight (angle of the sun's rays) decreases with increasing latitude. Each latitude thus presents a different combination of day-length and maximum sunlight intensity.

11. *Fifth Annual Report*, Iowa, 1858; 8.

12. Richard Steckel, "The Economic Foundations of East-West Migration during the Nineteenth Century," *Explorations in Economic History* 20 (1982), 14–36.

13. Will, *Corn for the Northwest*, 87. Many later developments in the Upper Middle

West are described in Herbert Kendall Hayes, *A Professor's Story of Hybrid Corn* (Minneapolis: Burgess, 1963).

14. Howard G. Roepke, "Changes in Corn Production on the Northern Margins of the Corn Belt," *Agricultural History* 33 (1959), 126–32; and Andreas Grotewold, *Regional Changes in Crop Production in the United States from 1909 to 1949*, research paper no. 40 (Chicago: University of Chicago, Department of Geography, 1955).

15. E. Z. Russell et al., "Hog Production and Marketing," *Yearbook, 1922* (Washington, D.C.: U.S. Department of Agriculture, 1922); fig. 30 shows receipts of hogs and public stockyards, 1916–1921.

16. J. K. Hudson, "Essay on Grains," *Transactions of the Kansas State Board of Agriculture*, 1872; 276.

17. *U.S. Census*, 1900, Agriculture, plate 7.

18. Nathaniel S. Shaler, *Kentucky: A Pioneer Commonwealth* (Boston: Houghton, Mifflin, 1888), 29.

19. Changes in the Nashville Basin are described in Stephen V. Ash, *Middle Tennessee: Society Transformed, 1860–1870* (Baton Rouge: Louisiana State University Press, 1988), 188.

20. Claypan soils, also formerly known as planosols, are now generally classed as Albaqualfs; see Henry D. Foth and John W. Schaefer, *Soils Geography and Land Use* (New York: Wiley, 1980), 161–65. Their inferior qualities for corn meant the Corn Belt had only a brief season in the southern Illinois gray lands; James W. Whitaker, *Feedlot Empire: Beef Cattle Feeding in Illinois and Iowa, 1840–1900* (Ames: Iowa State University Press, 1975), 77. Planosols in southern Iowa account for an area of relatively low agricultural land values within the state; Neil E. Salisbury, "Agricultural Productivity and the Physical Resource Base of Iowa, *Iowa Business Digest* 31, no. 2 (1960), 27–31.

21. M. E. Ensminger, *Swine Science* (Danville, Ill.: Interstate, 1961), 53.

22. Alonzo E. Taylor, *Corn and Hog Surplus of the Corn Belt* (Stanford: Stanford Food Research Institute, 1932), 88–92, 125.

23. U.S. Bureau of the Census, *Historical Statistics of the United States, Colonial Times to 1957* (Washington, D.C., 1960), Series K. The history of this period in American agriculture is covered in James H. Shideler, *Farm Crisis, 1919–1923* (Berkeley: University of California Press, 1957).

24. Taylor, *Corn and Hog Surplus*, 125. The basis for the corn-hog ratio is explained in Henry A. Wallace, *Agricultural Prices* (Des Moines: Wallace, 1920).

25. Theodore Saloutos and John D. Hicks, *Agricultural Discontent in the Middle West, 1900–1939* (Madison: University of Wisconsin Press, 1951), 486; Saloutos, *The American Farmer and the New Deal* (Ames: Iowa State University Press, 1982), 70–72; Chicago Drovers Journal, *Yearbook of Figures of the Livestock Trade*, 1953, 31.

26. Charles V. Piper and William J. Morse, *The Soybean* (New York: McGraw-Hill, 1923), v, 4.

27. Nelson Klose, *America's Crop Heritage: The History of Foreign Plant Introduction by the Federal Government* (Ames: Iowa State College Press, 1950), 14, 134–36; Piper and Morse, *The Soybean*, 40–41.

28. Edward Jerome Dies, *Soybeans: Gold from the Soil* (New York: Macmillan, 1944), 14–16; A. H. Probst and R. W. Judd, "Origin, U.S. History and Development, and World Distribution," in B. E. Caldwell, ed., *Soybeans: Improvement, Production, and Uses* (Madison: American Society of Agronomy, 1973), 1–15.

29. *Proceedings of the American Soybean Association* 1 (1925–1927), 9–10, 39–40.

30. Piper and Morse, *The Soybean*, 19, 219; *15th Proceedings*, American Soybean Assoc., 1935, 4.

31. R. A. Boyer, "How Soybeans Help Make Fords," *16th Proceedings*, Am. Soybean

Assoc., 1936, 6–9. See also W. B. Van Arsdel, "The Industrial Market for Farm Products," *Farmers in a Changing World: Yearbook of Agriculture, 1940* (Washington, D.C. U.S. Department of Agriculture, 1940), 606–26.

32. F. A. Wand, "Relations between the Soybean Grower and the Oil Mill," *Proceedings*, Am. Soybean Assoc., vol. 1, 104–105; Dies, *Soybeans*, 17; Leo G. Windish, *The Soybean Pioneers* (Galva, Ill.: by author, 1981), 67, chaps. 20–21; Deborah Fitzgerald, *The Business of Breeding: Hybrid Corn in Illinois, 1890–1940* (Ithaca: Cornell University Press, 1990), 117.

33. Windish, *Soybean Pioneers*, 65–67; Wand, *Proceedings*, American Soybean Assoc., vol. 1, 115.

34. *15th Proceedings*, Am. Soybean Assoc., 1935; Dies, *Soybeans*, 50–53; *20th Proceedings*, Am. Soybean Assoc., 1940, 76.

35. Windish, *Soybean Pioneers*, 82–86; *19th Proceedings*, Am. Soybean Assoc., 1939, 10–18; Dies, *Soybeans*, 27–29.

36. News items and advertisements in *Proceedings*, Am. Soybean Assoc., 1935–40. For a survey of recent distribution patterns see W. J. Free, "Location, Type, and Size of the U.S. Soybean Processing Industry," in Lowell C. Hill, ed., *World Soybean Research* (Danville, Ill.: Interstate, 1974), 715–23. According to Free, soybeans should be crushed in the area where they are consumed.

37. Earl C. Hedlund, *The Transportation Economics of the Soybean Processing Industry*, Illinois Studies in the Social Sciences, vol. 33, no. 1 (Urbana: University of Illinois Press, 1952), chaps. 5–7.

38. Merle C. Prunty, Jr., "Soybeans in the Lower Mississippi Valley," *Economic Geography* 26 (1950), 301–14. For a general overview of early soybean production and processing see Alvin A. Munn, "Production and Utilization of the Soybean in the United States," *Economic Geography* 26 (1950), 223–34.

39. E. T. Delwiche, "Extending the Soybean Belt Northward," *19th Proceedings*, Am. Soybean Assoc., 1939, 22, reported that soybeans were grown experimentally on former Norway- and jack-pine land near Iron River and Spooner, Wisconsin, between 1906 and 1910.

40. Production figures reported in Dies, *Soybeans*, 25, and *Soybean Digest Bluebook* (Hudson, Ia.: Am. Soybean Assoc., 1971).

41. A. Richard Crabb, *The Hybrid-Corn Makers: Prophets of Plenty* (New Brunswick, N.J.: Rutgers University Press, 1947), 144–45; Russell Lord, *The Wallaces of Iowa* (Boston: Houghton-Mifflin, 1947), 140–59.

42. Werner Mehrle, "Betriebswirtschaftliches und Technisches Zum Maisanbau in den U.S.A.," doctoral dissertation, Technischen Hochschule Munchen, 1957, shows graphs of corn yield in Illinois, 1865 to 1950. Also see Andreas Grotewold, "Changing Patterns of Corn Yields per Acre: The Position of Illinois in the United States," *Transactions*, Illinois Academy of Science, 48 (1956), 157–65.

43. Martin L. Mosher, *Early Iowa Corn Yield Tests and Related Later Programs* (Ames: Iowa State University Press, 1962), 75–89.

44. Edgar Anderson and William L. Brown, "Origin of Corn Belt Maize and Its Genetic Significance," John W. Gowen, ed., *Heterosis* (Ames: Iowa State University Press, 1952), 124–48. The Flint-Dent crosses (Corn Belt Dents) were hybrids that, in succeeding generations, backcrossed to one or both parents. Anderson used the term "introgression" to describe the "gradual infiltration of the germ plasm of one species into that of another"; Edgar Anderson, *Introgressive Hybridization* (New York: Wiley, 1949), 1. "As cultivation of maize became more extensive, introgression among the selected strains occurred because isolation was reduced. The amount of introgression . . . depended on the topography of the areas and the amount of interchange of seed among pioneer settlers"; Arnel L. Hallauer

and J. B. Miranda, Fo. *Quantitative Genetics in Maize Breeding* (Ames: Iowa State University Press, 1981), 4.

45. Crabb, *Hybrid-Corn Makers*, 45–56.

46. Mosher, *Corn Yield Tests*, 85–89.

47. Helen M. Cavanagh, *Seed, Soil and Science: The Story of Eugene D. Funk* (Chicago: Lakeside Press, 1959), 85. Also see M. L. Bowman and B. W. Crossley, *Corn Growing, Judging, Breeding, Feeding, Marketing* (Ames, Ia.: by authors, 1911), 3. John S. Leaming, *Corn and Its Culture, by a Pioneer Corn Raiser with 60 Years Experience in the Cornfield* (Wilmington, Oh.: Journal Steam Print, 1883), claimed a date of 1826 for the origin of his corn; this was in some doubt, however; see H. A Wallace and E. N. Bressman, *Corn and Corn Growing* (Des Moines: Wallace, 1925), 181.

48. Cavanagh, *Eugene D. Funk*, map facing p. 5.; Helen M. Cavanagh, *Funk of Funk's Grove: Farmer, Legislator, and Cattle King of the Old Northwest, 1797–1865* (Bloomington, Ill.: Pantagraph, 1952), chap. 1.

49. Cavanagh, *Eugene D. Funk*, 69.

50. Cavanagh, *Eugene D. Funk*, 326; Holbert's role is stressed throughout Crabb, *Hybrid-Corn Makers*.

51. Edward L. Schapsmeier and Frederick H. Schapsmeier, *Henry A. Wallace of Iowa: The Agrarian Years, 1910–1940* (Ames: Iowa State University Press, 1968), 27; Fitzgerald, *Business of Breeding*, 161; Cavanagh, *Eugene D. Funk*, 380–83; Crabb, *Hybrid-Corn Makers*, 151–54.

52. Crabb, *Hybrid-Corn Makers*, xii, 84–86. Also see John M. Airy, "Production of Hybrid Corn Seed," in George F. Sprague, ed., *Corn and Corn Improvement* (New York: Academic, 1955), 379–422.

53. Fitzgerald, *Business of Breeding*, 129.

54. Zvi Griliches, "Hybrid Corn: An Exploration in the Economics of Technological Change," *Econometrica* 25 (1957), 501–23. Griliches used the logistic function to predict $p(t)$, the proportion of adopters at time $t = K/\{1 + \exp[-(a + bt)]\}$ where K is interpreted as the "ceiling," or maximum percentage adopters, a is the origin, and b is the rate of adoption. The time of ten percent adoption is $t = -(a + 2.2)/b$; the time of fifty percent adoption $= \{-[\log_e(K - .5)/.5] + a\}/b$. The control values for map isopleths were obtained by substituting into the equation the numbers Griliches lists for the crop reporting districts.

55. Lawrence A. Brown, *Innovation Diffusion: A New Perspective* (London: Methuen, 1981), fig. 1.3; Brown also summarizes issues in the exchange between Griliches and rural sociologists who believed Griliches's economic explanation of hybrid corn adoption did not take social factors into account; see E. M. Rogers and E. Havens, "Adoption of Hybrid Corn: Profitability and the Interaction Effect," *Rural Sociology* 26 (1961), 409–14; Griliches, "Profitability versus Interaction: Another False Dichotomy," *Rural Sociology* 27 (1962), 327–30; and Rogers and Havens, "Rejoinder to Griliches' 'Another False Dichotomy,'" *Rural Sociology* 27 (1962), 330–32. G. F. Sprague, "Development and Adoption of Hybrid Corn," *Illinois Research*, vol. 34, nos. 1–2 (1992), 15–18, provides an overview of the trends in hybrid corn production.

56. Griliches, "Hybrid Corn," 519.

57. Crabb, *Hybrid-Corn Makers*, 291.

58. Dorothy Giles, *Singing Valleys: The Story of Corn* (New York: Random House, 1940); Saloutos, *New Deal*, 72–75; Fitzgerald, *Business of Breeding*, 196.

59. Harold Lee, *Roswell Garst: A Biography* (Ames: Iowa State University Press, 1984), 120–24, describes one well-known farmer's reactions to government acreage controls on corn. Other impacts of federal farm policies are covered in Michael W. Schuyler, *The Dread*

of Plenty: Agricultural Relief Activities of the Federal Government in the Middle West, 1933–1939 (Manhattan, Kans.: Sunflower University Press, 1989).

60. Willard W. Cochrane and Mary E. Ryan, *American Farm Policy, 1948–1973* (Minneapolis: University of Minnesota Press, 1976), 176–82.

11. West to the Plains

1. Edward A. Duddy and David A. Revzan, "The Changing Relative Importance of the Central Livestock Market," *Journal of Business of the University of Chicago*, vol. 11, no. 3 (1938), chap. 2. St. Joseph is out of sequence geographically; its stockyard was a later addition, squeezed in between Omaha and Kansas City in a major livestock-producing area. The early industry at Kansas City is described in George T. Renner, "The Kansas City Meat Packing Industry before 1900," *Missouri Historical Review* 55 (1960), 18–29.

2. Figures are taken from summations in the annual, *Yearbook of Figures of the Livestock Trade* (Chicago: Drovers Journal). Also see Howard C. Hill, "The Development of Chicago as a Center of the Meat Packing Industry," *Mississippi Valley Historical Review* 10 (1923), 253–73.

3. Swift and Company, *Annual Report*, 1912; 16–17. The illustration assumed that steers purchased at $4.60 cwt at Ft. Worth were dressed and sold at $6.583 cwt, a loss given that 44% of the weight was outside the meat. But the hides, hooves, horns, and remainder, converted into lard, oleomargarine, soap, washing powder, cleanser, and glue (all bearing Swift product names) earned a profit.

4. Duddy and Revzan, "Central Livestock Market," chaps. 1, 2; Robert Aduddell and Louis Cain, "Location and Collusion in the Meat Packing Industry," in Louis P. Cain and Paul J. Uselding, eds., *Business Enterprise and Economic Change: Essays in Honor of Harold F. Williamson* (Kent: Kent State University Press, 1973), 106–109.

5. Geo. A. Hormel and Company, *Financial Report*, 1967, 16. The arrival of meat packing in Iowa's second-rank cities is discussed in Harold H. McCarty and C. Woody Thompson, *Meat Packing in Iowa*, Iowa Studies in Business, vol. 12 (Iowa City: University of Iowa Bureau of Business and Economic Research, 1933).

6. Duddy and Revzan, "Central Livestock Market," 48–59.

7. Robert Mitchell Aduddell, "The Meat Packing Industry and the Consent Decree, 1920–1956," Ph.D. dissertation, Northwestern University, 1971, table 5.

8. Fig. 46 is based on hogs and pigs sold alive, by county (*Census of Agriculture*, 1954); number of hogs slaughtered, by state (*Census of Manufactures*, 1954); and hog receipts and hogs slaughtered at 65 markets (*Drovers Journal Yearbook*, 1954–1955). Flow lines are drawn such that the reported numbers of hogs arriving in terminal markets and the numbers reported slaughtered make a surplus that moves to the nearest outside market. Volume at interior packers (not included in the 65) is estimated by apportioning the residual of state totals so as to minimize movement of live animals.

9. Swift and Company, *Annual Report*, 1959, 1965–1972; Esmark, *Annual Report*, 1972–1975. Also see Richard J. Arnould, "Changing Patterns of Concentration in American Meat Packing, 1880–1963," *Business History Review* 45 (1971), 19–34.

10. Wilson and Company, *Annual Report*, 1967–1972; Armour and Company, *Annual Report*, 1951, 1960–1967; *The Greyhound Corporation Annual Report*, 1972–1974; ConAgra, *Annual Report*, 1972; 1980–1987.

11. Eugene Mather, "The Production and Marketing of Wyoming Beef Cattle," *Economic Geography* 26 (1950), 81–93.

12. Many range cattle came to Chicago for redistribution; see Edward A. Duddy and David A. Revzan, *The Distribution of Livestock from the Chicago Market, 1924–29*, Studies in Business Administration, vol. 3, no. 1 (Chicago: University of Chicago Press, 1932), fig. 1. Also see Duddy and Revzan, *The Supply Area of the Chicago Livestock Market*, Studies in Business Administration, vol. 2, no. 1 (Chicago: University of Chicago Press, 1931).

13. Mather, "Wyoming Beef Cattle," fig. 14; and Robert A. Kennelly, "Cattle-Feeding in the Imperial Valley," Association of Pacific Coast Geographers *Yearbook*, 1960, 50–56. See also H. Bowman Hawkes, "Livestock Industry," in Clifford M. Zierer, ed., *California and the Southwest* (New York: Wiley, 1956), 176–91; Herrell DeGraff, *Beef Production and Distribution* (Norman: University of Oklahoma Press, 1960); John T. Schlebecker, *Cattle Raising on the Plains, 1900–1961* (Lincoln: University of Nebraska Press, 1963); and L. F. Schrader and G. A. King, "Regional Location of Beef Cattle Feeding," *Journal of Farm Economics* 44 (1962), 64–81.

14. Kennelly, "Imperial Valley," figs. 1, 2.

15. John Fraser Hart, "Meat, Milk, and Manure Management in the West," *Geographical Review* 54 (1964), 118–19; Charles Bussing, "The Impact of Feedlots," in Donald D. MacPhail, ed., *The High Plains: Problems of Semiarid Environments*, Contribution 15, Committee on Desert and Arid Zones Research, Southwestern and Rocky Mountain Division, A. A. A. S. (Fort Collins: Colorado State University, 1972), 84–85; and D. L. Wheeler, "The Cattle Feeding Industry in the Southern High Plains of the United States," *Geography* 64 (1979), 50–51.

16. Garry L. Nall, "The Cattle Feeding Industry on the Texas High Plains," Henry C. Dethloff and Irvin M. May, Jr., eds., *Southwestern Agriculture: Pre-Columbian to Modern* (College Station: Texas A&M University Press, 1982), 109.

17. Richard E. Stevens, "Cattle Feeding on the Llano Estacado," *University of Colorado Studies*, Series in the Earth Sciences, vol. 3 (1965), 61–76.

18. Several "bicentennial" histories of Texas Panhandle counties reveal the presence of early Illinoisians and other middle-westerners who came to Texas, perhaps at the urging of railroads and land companies. See *A History of Parmer County, Texas* (Bovina, Tex.: Parmer County Historical Society, 1974), 3; and *A History of Crosby County, 1876–1977* (Crosbyton, Tex.: Crosby County Historical Commission, 1978), 451–52. See also Jo Stewart Randel, *A Time to Purpose: A Chronicle of Carson County and Area*, 4 vols. (Hereford, Tex.: Carson County Square House Museum, 1972); and *God, Grass & Grit: History of the Sherman County Trade Area*, 2 vols. (Seagraves, Tex.: Pioneer, 1975), passim. An excellent overview of early settlement and subsequent irrigation in the Texas Panhandle is Bret Wallach, *At Odds with Progress* (Tucson: University of Arizona Press, 1991), chap. 8.

19. Stevens, "Llano Estacado," 64.

20. Jack R. Harlan, *Crops and Man* (Madison: American Society of Agronomy, 1975), 92, 117; Hugh Doggett, *Sorghum*, 2d ed. (London: Longman Scientific, 1988), chap. 1.

21. Nelson Klose, *America's Crop Heritage: The History of Foreign Plant Introduction by the Federal Government* (Ames: Iowa State College Press, 1950), 47–50; Doggett, *Sorghum*, 65.

22. J. Roy Quinby, "Hybrid Sorghum: A Triumph of Research," in Dethloff and May, eds., *Southwestern Agriculture*, 93–105; Klose, *America's Crop Heritage*, 49.

23. Quinby, "Hybrid Sorghum," 96; K. F. Schertz, "Sorghum Research in the U.S.A.," *Report of the 23rd Annual Corn and Sorghum Research Conference*, 1968, 110–19.

24. Stevens, "Llano Estacado," 64–65; Nall, "Texas High Plains," 110.

25. In Kansas, where natural gas is available to fuel irrigation pumps, the extra costs of irrigating corn were low and far more than offset by corn's greater yield and higher

price; see "Irrigation Fuel Costs," *Energy Facts*, MF-501 (Manhattan: Kansas State University, Cooperative Extension Service, 1979).

26. Charles E. Bussing, "The Cattle Feeding Industry of the Northern Colorado Piedmont—A Geographical Appraisal," Ph.D. dissertation, University of Nebraska, Lincoln, 1968, 42–48.

27. Henry Hart, *The Dark Missouri* (Madison: University of Wisconsin Press, 1957), 156–57; Bussing, "Northern Colorado Piedmont," fig. 16.

28. David A. Henderson, " 'Corn Belt' Cattle Feeding in Eastern Colorado's Irrigated Valleys," *Economic Geography* 30 (1954), 364–72.

29. Harold Lee, *Roswell Garst: A Biography* (Ames: Iowa State University Press, 1984), 70–79.

30. Irrigation in the Arkansas Valley is the subject of James Earl Sherow, *Watering the Valley: Development along the High Plains Arkansas River, 1870–1950* (Lawrence: University Press of Kansas, 1990). For the Loup Valley see Richard G. Bremer, *Agricultural Change in an Urban Age: The Loup Country of Nebraska, 1910–1970*, University of Nebraska Studies, n.s., no. 51 (Lincoln: University of Nebraska Press, 1976). Henry A. Wallace vigorously embraced irrigation for the Plains, as shown in his "On the Trail of the Corn Belt Farmer," in *Henry A. Wallace's Irrigation Frontier*, Richard Lowitt and Judith Fabry, eds. (Norman: University of Oklahoma Press, 1991), 20, 26–27.

31. Charles Bowden, *Killing the Hidden Waters* (Austin: University of Texas Press, 1977), 99–125.

32. John B. Weeks and Edwin D. Gutentag, "Region 17, High Plains," in William Back, Joseph S. Rosenshein, and Paul R. Seaber, eds., *Hydrogeology, The Geology of North America*, vols. 0–2 (Boulder, Colo.: Geological Society of America, 1988), 157–64. See also William D. Thornbury, *Regional Geomorphology of the United States* (New York: Wiley, 1965), 300–302.

33. Leonard W. Bowden, *Diffusion of the Decision to Irrigate: Simulation of the Spread of a New Resource Management Practice in the Colorado Northern High Plains*, Research Paper 97 (Chicago: University of Chicago, Department of Geography, 1965), chap. 1.

34. C. Langdon White, Edwin J. Foscue, and Tom L. McKnight, *Regional Geography of Anglo-America*, 5th ed. (Englewood Cliffs, N.J.: Prentice-Hall, 1979), 340–41. Also see Tom L. McKnight, "Great Circles on the Great Plains," *Erdkunde* 33 (1979), 70–79.

35. Nall, "Texas High Plains," 112–14.

36. Nall, "Texas High Plains," 110; Stevens, "Llano Estacado," 75.

37. A "feed-grain county" is defined for this purpose as one producing a total of more than one million bushels of corn or sorghum in the High Plains section of Colorado, Kansas, New Mexico, Oklahoma, or Texas.

38. Bowden, *Decision to Irrigate*, 24.

39. Weeks and Gutentag, "High Plains," 160–62. The history of irrigation here is developed in Donald E. Green, *Land of the Underground Rain: Irrigation on the Texas High Plains, 1910–1970* (Austin: University of Texas Press, 1973).

40. Iowa Beef Packers, *Annual Report*, 1961, 1965; Iowa Beef Processors, *Annual Report*, 1979.

41. Iowa Beef Packers, *Annual Report*, 1965, 10.

42. Jimmy M. Skaggs, *Prime Cut: Livestock Raising and Meat Packing in the United States, 1607–1983* (College Station: Texas A&M University Press, 1986), 193.

43. Aduddell and Cain, "Meat Packing Industry," 101.

44. Bussing, "Northern Colorado Piedmont," 142–43; Monfort, *Annual Report*, 1979, 1981.

45. Bussing, "Northern Colorado Piedmont," 116–30; Monfort, *Annual Report*, 1981, 14.

46. Monfort, *Annual Report*, 1979, 1982, 1986; ConAgra, *Annual Report*, 1987.

47. Occidental Petroleum, *Annual Report*, 1987, 1990; IBP, *Annual Report*, 1991.

48. Skaggs, *Prime Cut*, 192–93. Chaps. 6 and 7 give an excellent overview of corporate strategies in the beef business since 1960.

49. Upton Sinclair, *The Jungle* [1906] (New York: New American Library, 1960). A perspective on the changes wrought in Garden City, Kansas, after Iowa Beef built their enormous Finney County processing plant in nearby Holcomb is found in Holly Hope, *Garden City: Dreams in a Kansas Town* (Norman: University of Oklahoma Press, 1988). Reactions to the present meat/livestock industry include Wayne Swanson and George Schultz, *Prime Rip* (Englewood Cliffs, N.J.: Prentice-Hall, 1982); and Jim Mason and Peter Singer, *Animal Factories*, 2d ed. (New York: Harmony, 1990).

50. Michael J. Broadway and Terry Ward, "Recent Changes in the Structure and Localization of the U.S. Meatpacking Industry," *Geography* 75 (1990), 76–79.

12. The Corn Business

1. Studies of middle-western farm tenancy have been largely concerned with testing Paul W. Gates's notion that tenancy grew out of flaws in United States land policies. See the articles in his *Landlords and Tenants on the Prairie Frontier: Studies in American Land Policy* (Ithaca: Cornell University Press, 1963). Donald L. Winters, *Farmers without Farms: Agricultural Tenancy in Nineteenth-Century Iowa* (Westport, Conn.: Greenwood Press, 1978), 106–107, concludes, however, that the factors Gates identified "had little if anything to do with the incidence of farm renting in Iowa." Also on Iowa see Seddie Cogwell, Jr., *Tenure, Nativity, and Age as Factors in Iowa Agriculture, 1850–1880* (Ames: Iowa State University Press, 1975). Jeremy Atack, "Tenants and Yeomen in the Nineteenth Century," *Agricultural History* 62 (1988), 32, regarded tenancy in the latter half of the nineteenth century as "an evolutionary, rather than revolutionary, process." Studies of specific individuals have shown that landlords did not have the negative effect sometimes supposed; see Allan Bogue, "Foreclosure Tenancy on the Northern Plains," *Agricultural History* 39 (1965), 3–16; and Homer Socolofsky, *Landlord William Scully* (Lawrence: Regents Press of Kansas, 1979).

2. *Thirteenth Census of the United States*, 5, Agriculture, 100–120.

3. On the Des Moines lobe see Leslie Hewes and Philip E. Frandson, "Occupying the Wet Prairie: The Role of Artificial Drainage in Story County, Iowa," *Annals*, Association of American Geographers, 42 (1952), 24–50.

4. Theodore Saloutos and John D. Hicks, *Agricultural Discontent in the Middle West, 1900–1939* (Madison: University of Wisconsin Press, 1951), 101–103.

5. Those who failed to attain farm ownership also added to the westward movement. In Sangamon County, Illinois, in the 1840s, those who owned land were far more likely to remain than were those who owned no land; John Mack Farragher, "Open-Country Community, Sugar Creek, Illinois, 1820–1850," in Steven Hahn and Jonathan Prude, eds., *The Countryside in the Age of Capitalist Transformation: Essays in the Social History of Rural America* (Chapel Hill: University of North Carolina Press, 1985), 233–58.

6. United States Special Committee on Farm Tenancy, *Report of the President's Committee on Farm Tenancy* (Washington, D.C.: U.S. Natural Resources Committee, 1937); Paul V. Maris, "Farm Tenancy," *Farmers in a Changing World*, Yearbook of Agriculture, 1940 (Washington, D.C.: U.S. Department of Agriculture, 1940), 887–906. Federal and state legislation proposed in the late 1930s included differential taxes on larger holdings, deferral

of foreclosure action, and increased capital-gains taxes on sales of land. The Bankhead-Jones Farm Tenancy Act of 1937 established loans to allow families to purchase farms. The Roosevelt administration's view of farm tenancy is vividly illustrated in the full-page tableau "The Agricultural Ladder," in which laborers climb to sharecropping, then to tenant farming, and finally become owner-operators; Raymond C. Smith, "New Conditions Demand New Opportunities," *Farmers in a Changing World*, fig. 1.

7. United States Tariff Commission, *Corn or Maize: Differences in Costs of Production of Corn or Maize in the United States and in the Principal Competing Country* (Washington, D.C.: U.S. Government Printing Office, 1929), 13–25.

8. Harold Lee, *Roswell Garst: A Biography* (Ames: Iowa State University Press, 1984), 45.

9. John Fraser Hart, "Change in the Corn Belt," *Geographical Review* 76 (1986), 51–72; figs. 1, 4, 7. Papers describing the shift in technology and its impact on the farm include Allan Bogue, "Changes in Mechanical and Plant Technology: Corn Belt, 1910–1940," *Journal of Economic History* 443 (1983), 25–26; Sam B. Hilliard, "The Dynamics of Power and Technology: Recent Trends in Mechanization on the American Farm," *Technology and Culture* 13 (1972), 1–24; Everett G. Smith, Jr., "Fragmented Farms in the United States," *Annals*, Association of American Geographers, 65 (1975), 58–70; and Hiram M. Drache, "Midwest Agriculture: Changing with Technology," in Vivian Wiser, ed., *Two Centuries of American Agriculture* (Washington, D.C.: Agricultural History Society, 1976), 290–302. Also see R. Douglas Hurt, "Ohio Agriculture since World War II," *Ohio History* 97 (1988), 50–71.

10. Mildred Smith Finney, "The Corn Refining Industries: A Study in Industrial Location," Ph.D. dissertation, Northwestern University, 1959, chap. 1.

11. Finney, "Corn Refining," tables 1, 5; *The Story of Corn and Its Products* (New York: Corn Industries Research Foundation, 1952).

12. Finney, "Corn Refining," 144–45; for a general discussion of the transit privilege and its impact on agricultural industries see Earl C. Hedlund, *The Transportation Economics of the Soybean Processing Industry*, Illinois Studies in the Social Sciences, vol. 33, no. 1 (Urbana: University of Illinois Press, 1952), chap. 5.

13. Frederick L. Jeffries, "Grinding Corn," Corn Refiners Association, *Corn Annual*, 1985, 18.

14. H. A. Bendixen, "High Fructose Corn Syrup," *Corn Annual*, 1973, 11–13. An excellent account of corn's place in the various new biotechnologies is found in Betty Fussell, *The Story of Corn* (New York: Knopf, 1992), 265–78.

15. *Corn Annual*, 1976, 24; 1978, 17.

16. Archer Daniels Midland, *Annual Report*, 1985, 6.

17. J. W. Dudley, ed., *Seventy Generations of Selection for Oil and Protein in Maize* (Madison: Crops Science Society of America, 1974), 211.

18. E. T. Mertz et al., "Better Protein Quality in Maize," *Advances in Chemistry* 57 (1966), 228–42.

19. Earl Butz, "Annual Letter," *Corn Annual*, 1975, 6; William E. Bangs and Gary A. Reineccius, "Corn Starch Derivatives: Possible Wall Materials for Spray-Dried Flavor Manufacture," in Sara J. Risch and Gary A. Reineccius, eds., *Flavor Encapsulation* (Washington, D.C.: American Chemical Society, 1988), 12–28.

20. V. Daniel Hunt, *The Gasohol Handbook* (New York: Industrial Press, 1981), 1–3, 514. Also see Dwight L. Miller, "The Corn Belt—An Energy Belt?" *Corn Annual*, 1979, 26.

21. Under present (1992–93) law, gasohol is exempt from 6 of the 9 cents in federal excise taxes on gasoline. Byproducts of ethanol manufacture include corn oil as well as

materials that can be manufactured into livestock feed. The tax subsidy increases the demand for corn and increases its price. The byproducts form stocks of goods that might otherwise be manufactured from soybeans. Thus while gasohol meets environmental concerns, it has also increased food prices by a small amount. Apparently ethanol production remains uneconomical without the subsidy. See "Ethanol Could Affect Corn Prices, Farm Income, and Government Outlays," U.S. Department of Agriculture, *Agricultural Outlook*, July 1988, 28. Some of the problems and politics are presented in United States General Accounting Office, *Alcohol Fuels: Impacts from Increased Use of Ethanol Blended Fuels*, Report to the Chairman, Subcommittee on Energy and Power, Committee on Energy and Commerce, U.S. House of Representatives, 101st Congress (Washington, D.C., 1990); and *Review of the Role of Ethanol in the 1990s*, Joint Hearing before the Subcommittee on Forests, Family Farms, and Energy and the Subcommittee on Wheat, Soybean, and Feed Grains of the Committee on Agriculture and the Subcommittee on Energy and Power of the Committee on Energy and Commerce, U.S. House of Representatives, 100th Congress (Washington, D.C.: 1988).

22. *Report of the Commissioner of Patents*, 1849, part 2, Agriculture (Washington, D.C., 1850), 18. Gershom Flagg, the Vermonter who settled at Edwardsville, Illinois, wrote to his brother in Vermont in 1818, "The method of raising Corn here is to plough the ground once then furrow it both ways and plant the Corn 4 feet each way and plow between it 3 or 4 times but never hoe it at all"; letter to Artemas Flagg, September 12, 1818, in Barbara Lawrence and Nedra Branz, *The Flagg Correspondence: Selected Letters, 1816–1854* (Carbondale: Southern Illinois University Press, 1986), 19. Despite technological progress, this was the basic culture used through the 1950s.

23. *Commissioner of Patents*, 1849, 18–19.

24. W. L. Brown, "Physical Characteristics of Corn of the Future," *Proceedings of the Twentieth Annual Hybrid Corn Industry-Research Conference*, 1965, 7–16.

25. Statement of John Tarr, Fayette County, Pennsylvania, in *Commissioner of Patents*, 1849, 19–20. Numerous modern counterparts of the early rotations are discussed in John Fraser Hart, *The Land That Feeds Us* (New York: Norton, 1991).

26. H. A. Wallace and E. N. Bressman, *Corn and Corn Growing* (Des Moines: Wallace, 1925), 24.

27. Arnel L. Hallauer and J. B. Miranda, Fo., *Quantitative Genetics in Maize Breeding* (Ames: Iowa State University Press, 1981), 7; John S. Rogers, "Use of Male-Sterile Stocks in the Production of Corn Hybrids," *Proceedings of the Seventh Annual Hybrid Corn-Industry Research Conference*, 1952, 7–15; and Helen M. Cavanagh, *Seed, Soil and Science: The Story of Eugene D. Funk* (Chicago: Lakeside Press, 1959), 428.

28. C. C. Cockerham, "Implications of Genetic Variances in a Hybrid Breeding Program," *Crop Science* 1 (1961), 47–51; Hallauer and Miranda, *Quantitative Genetics*, 8.

29. Hallauer and Miranda, *Quantitative Genetics*, 9.

30. Brown, "Physical Characteristics of Corn," 7–16. Historical trends in corn yields are illustrated in G. F. Sprague, "Development and Adoption of Hybrid Corn," *Illinois Research*, vol. 34, nos. 1–2 (1992), 15–18.

31. As on the soybean farms of Benton County, Indiana, for example; see John Fraser Hart, "Field Patterns in Indiana," *Geographical Review* 58 (1968), 450–71.

32. Brown, "Physical Characteristics of Corn," 10.

33. Earl L. Butz, "Corn Has a Bright Future in America," *Corn Annual*, 1973, 5–7. For a concise view of Butz's stewardship as Secretary of Agriculture see [Lauren Soth], *The Farm Policy Game, Play by Play* (Ames: Iowa State University Press, 1989), 2–4. Conditions leading up to the grain export boom are discussed in Harry D. Fornari, "U.S. Grain Exports: A Bicentennial Overview," in Vivian Wiser, ed., *Two Centuries of American Agriculture* (Washington, D.C.: Agricultural History Society, 1976), 137–50.

34. Earl L. Butz, "Corn Holds the Key to Affluence," *Corn Annual*, 1976, 9. On changes in agricultural policy see Willard W. Cochrane and Mary E. Ryan, *American Farm Policy, 1948–1973* (Minneapolis: University of Minnesota Press, 1976), 179.

35. John C. Hudson, "New Grain Networks in an Old Urban System," in John Fraser Hart, ed., *Our Changing Cities* (Baltimore: Johns Hopkins University Press, 1991), 86–107. Grain flows under domestic versus foreign demand conditions are analyzed in D. A. Hilger, B. A. McCarl, and J. W. Uhrig, "Facilities Locations: The Case of Grain Subterminals," *American Journal of Agricultural Economics* 59 (1977), 674–82.

36. National Waterways Foundation, *U.S. Waterways Productivity* (Huntsville, Ala.: Strode, 1983), 57. The revival of water transportation in corn exports is vividly illustrated in Darrell L. Good and Lowell D. Hill, "Illinois and the World Corn Market," *Illinois Research*, vol. 34, nos. 1–2 (1992), 3–7.

37. Mark Friedberger, *Farm Families and Change in Twentieth Century America* (Lexington: University Press of Kentucky, 1988), 191–93.

38. Neil E. Harl, *The Farm Debt Crisis of the 1980s* (Ames: Iowa State University Press, 1990), 13–17, identifies the inflationary economy, the Federal Reserve's tightening of the credit supply in 1979, and federal budget deficits traceable to the 1981 tax cuts as major contributing factors to the farm debt crisis from outside the agricultural sector. Also see *The Iowa Economy* (Chicago: Federal Reserve Bank of Chicago, 1987).

39. Friedberger, *Farm Families*, 221. The protest literature includes Jim Schwab, *Raising Less Corn and More Hell: Midwestern Farmers Speak Out*, foreword by Senator Tom Harkin (Urbana: University of Illinois Press, 1988); and Paul C. Rosenblatt, *Farming Is in Our Blood: Farm Families in Economic Crisis* (Ames: Iowa State University Press, 1990).

40. U.S. Department of Agriculture, *Agricultural Statistics, 1987* (Washington, D.C., 1987), table 1.

41. *Agricultural Outlook*, September 1991, 31; March 1990, 19; March 1986, 24.

42. *Agricultural Outlook*, June 1992, 20–21.

43. *Agricultural Outlook*, September 1991, 22; U.S. Department of Agriculture, *Agricultural Statistics*, 1990, table 38. The overall industry conditions were assessed in Mack N. Leath, Lynn H. Meyer, and Lowell D. Hill, *U.S. Corn Industry*, Agricultural Economic Report 479 (Washington, D.C.: U.S. Department of Agriculture, 1982).

44. James M. McDonald, *Effects of Railroad Deregulation on Grain Transportation*, Economic Research Service, Technical Bulletin 1759 (Washington, D.C.: U.S. Department of Agriculture, June 1989).

45. *Agricultural Outlook*, December 1988, 23–24.

46. Cavanagh, *Eugene D. Funk*, 126.

47. G. Gajewski et al., "Sustainable Agriculture: What's It All About?" *Agricultural Outlook*, May 1992, 24–25. Also see Philip J. Gersmehl, "No-Till Farming: The Regional Applicability of a Revolutionary Agricultural Technology," *Geographical Review* 68 (1978), 66–79.

48. John T. Pierce, *The Food Resource* (London: Longman Scientific, 1990).

49. Ralph W. F. Hardy, "Biotechnology and New Directions for Agrochemicals," in Gino R. Marco, Robert M. Hollingsworth, and Jack R. Plimmer, eds., *Regulation of Agrochemicals: A Driving Force in Their Evolution* (Washington, D.C.: American Chemical Society, 1991), 131–44. Also see S. H. Mantell, J. A. Matthews, and R. A. McKee, *Principles of Plant Biotechnology: An Introduction to Genetic Engineering in Plants* (Oxford: Blackwell Scientific, 1985); and Gabrielle J. Persley, ed., *Beyond Mendel's Garden: Biotechnology in the Service of World Agriculture* (Oxford: CAB International, 1990).

50. Gustave K. Kohn, "Agriculture, Pesticides, and the American Chemical Industry," in Gino R. Marco, Robert M. Hollingsworth, and William Durham, eds., *Silent Spring Revisited* (Washington, D.C.: American Chemical Society, 1987), 159–79. Kohn enters the

claim (167–68) that the value of pesticides technology, fertilizer, and new seed varieties to the U.S. national wealth thus far has been $9.125 trillion.

51. Cynthia Rosenzweig, "Crop Response to Climatic Change in the Southern Great Plains: A Simulation Study," *Professional Geographer* 42 (1990), 20–37.

52. Karl B. Raitz and Carolyn Murray-Wooley, *The Gray and the Green: The Rock Fences of Kentucky's Bluegrass* (Lexington: University Press of Kentucky, 1990).

53. Bureau of the Census, *Residents of Farms and Rural Areas: 1989*, Current Population Reports, Series P-20, no. 446 (Washington, D.C., October, 1990).

Index

JOHN C. HUDSON is Professor of Geography at Northwestern University and the author of several books on the geography of the Middle West and Great Plains, including *Plains Country Towns* and *Crossing the Heartland*.